Y0-EGG-022

America against Poverty

America against Poverty

by Edward James

Lecturer in Social Administration
University of Birmingham

LONDON
ROUTLEDGE & KEGAN PAUL

First published 1970
by Routledge & Kegan Paul Ltd
Broadway House, 68-74 Carter Lane
London, E.C.4
Printed in Great Britain by
Northumberland Press Limited
Gateshead

SBN 7100 6760 7 (c)

Contents

General editor's introduction

The Library of Social Policy and Administration is designed to provide short texts suitable for the needs of Social Studies students in universities and other centres of higher education. They will also be of use to administrators in the social services, to practising social workers and to others whose work brings them into contact with the developing field of social service.

The Library will provide studies in depth rather than surveys of the whole field of social policy, and each volume will be complete in itself. Some will be studies of the British Social Services. Others will offer accounts of social policy in other countries, and provide material for comparative study. A third group will consist of case-studies in the processes of social policy.

For any student of social policy, the American Poverty War has been one of the major phenomena of the nineteen-sixties. It has spread and diversified with a speed which has confused observers in other countries and all but defied reasoned analysis. Those of us who have read Oscar Lewis and thought about the cultural poverty theory, followed the findings of Orshansky and Miller, reflected on Galbraith's *The Affluent Society*, Harrington's *Other America* and Marris and Rein's *Dilemmas of Social Reform*, still lack an overview of a decade of unprecedented change. For students who come fresh to the subject in the seventies, there is the problem of where to start, as the literature mounts and the areas of debate become increasingly complex.

Both groups may be grateful to Mr James, who writes this introductory account with clarity and scholarship. As a 'short-term foreign mercenary' in the Poverty War, he traces the multiple initiatives of the Kennedy-Johnson era with a lively eye for the significant points of change. If he is driven to the conclusion that 'Full employment, to which the federal government's main contribution was the Vietnam war, was the most potent poverty killer', he is aware also that irreversible changes have been initiated which may have long-term repercussions; and that some of these—the rediscovery of the existence of poverty in highly industrialized societies, the emphasis on community action and client participation—have major implications for the welfare societies of Western Europe as well as for the United States.

KATHLEEN JONES

Foreword

The War on Poverty was a phrase used by President Johnson to describe provisions of the Economic Opportunity Act of 1964. This is the sense in which I use the term in this book. However, the term is sometimes used to describe the entire anti-poverty efforts of the Kennedy-Johnson administration 1960-69, and I have been obliged to give a general outline of this legislation as well. The EOA measures are only understandable in this context.

Yet this book has to range even more widely. Any study in social policy in another nation must begin with the appreciation of the differences between that country and one's own. This book, then, first seeks to explain the United States. It could be argued that to explain a whole society is impossible except at immense length, destroying the usefulness of all comparative studies. However I believe that we can readily isolate some crucial differences that are relevant to social policy, and doing this will help us understand more of the interplay between social policy and its environment. In fact being overseas observers may give us certain insights that escape the natives, and which we ourselves perhaps lack in looking at our own society.

This is a personal perception of the War on Poverty

by one short-term foreign mercenary. More extensive and authoritative assessments are beginning to appear in America and I hope that my work will help bring them to a wider readership.

List of abbreviations for agencies and programmes

AB	Aid to the Blind
ADC	Aid to Dependent Children
AFDC	Aid to Families with Dependent Children
APTD	Aid to the Permanently and Totally Disabled
CAA	Community Action Agency
CAC	Community Action Council
CAMP	Central Area Motivation Program
CAP	Community Action Program
CEP	Concentrated Employment Program
EOA	Economic Opportunity Act
ESEA	Elementary and Secondary Education Act
FHA	Federal Housing Administration
GA	General Assistance
GEC	General Electric Company
HDC	Housing Development Corporations
HEW	Department of Health, Education and Welfare
HUD	Department of Housing and Urban Development
JC	Job Corps
JOBS	Job Opportunities in the Business Sector
MAA	Medical Aid for the Aged
MDTA	Manpower Development and Training Act
NIT	Negative Income Tax

LIST OF ABBREVIATIONS

NWRO	National Welfare Rights Organization
NYC	Neighborhood Youth Corps
OAA	Old Age Assistance
OASDHI	Old Age, Survivors, Disability and Health Insurance
OEO	Office of Economic Opportunity
OIC	Opportunities Industrialization Centers
PAAC	Philadelphia Anti-Poverty Action Committee
PACE	Public and Community Employment
PAE	Progress Aerospace Enterprises
PHDC	Philadelphia Housing Development Corporation
VA	Veterans' Administration
VISTA	Volunteers in Service to America

1

All Americans

What is the nearest foreign country to Britain? Forty London primary school children scribbled their answers. It was 1946 and theirs had been a makeshift education, sandwiched among air raids and evacuations, but most of them were confident on this question in the class test. Unhesitatingly fifteen children wrote—*America.*

Although the teacher disagreed, my fifteen former classmates represent a common British opinion. Clearly Britain and the United States have many things in common; a language (although the Americans have developed a slightly improved version), a legal system, a tradition of government, an illogical system of weights and measures, and much else, even to having the same nursery rhymes. So much of America is superficially so British that I feel I must start by emphasizing that America is different and cannot be judged in British terms. Britain is essentially a small European nation state. The USA is not.

The American identity

Starting with the obvious, the USA is big. Its main continental mass is as big as all Europe except for the Soviet Union and the range of climate and landscape is equally great. The distance from Miami in the south-east to Seattle in the north-west is 3,500 miles, equal to the distance between London and Jerusalem, and having driven the route in midsummer I feel it would do justice to any Crusade. So large a land cannot but be diverse. Yet the most striking feature of America, once one has grasped its size, is its degree of uniformity. Compared with travelling in Europe there is a release from all the problems of international frontiers, different languages and different currencies. The political achievement in America, in embracing this great sweep of territory and giving its 200 million inhabitants some sense of common national identity, is astonishing. That this not a strongly centralized nation and that it contains separatist elements is not nearly so remarkable as the fact that it ever came into being and that it survived.

Why does the USA exist? There is no national sentiment grounded on ethnic unity, on a common ancestral language, or a long tradition of living together in the same region which is the basis of European nationalism. The Americans are descendants of immigrants from all over Europe and much of Africa and Asia, who happened to settle in an almost empty continent a few generations ago. In its early history the various European colonial powers partitioned America as they were later to partition Africa, but in America the 'frontiers of colonialism' did not survive. The British colonies won their independence, and, much more surprisingly, set up a workable federation which within a century had spread across to the Pacific, swallowing most of the ex-French and Spanish territories. The federation survived its own internal tensions, which presented their greatest challenge in the Civil War, and lived to absorb floods of European migrants, mostly not of British descent. These people were acculturated, assimilated and taught to be 'Americans', integrated into the most successful consciously developed national sentiment of modern times.

America, like many new nations, has had to teach itself its national identity. The influx of migrants extended this task right into the twentieth century. The main instrument has been the school system, which has been free, public, and comprehensive for much longer and to a greater extent than in Britain, with a vigorous nationalism replacing the half-hearted religious element remaining in British schools. This self-conscious assertiveness with its devotion to flags and rituals may seem adolescent, but it is only since the Second World War that the overwhelming majority of the American population has been American-born of American-born parents. The War on Poverty is in some respects a further assertion of the American identity, assimilating America's 'last minority'. Poverty is un-American, and poverty groups need to be assimilated into the mainstream of American life, principally through the traditional path of education. It is ironic that this attempt at economic integration has sharpened the sense of racial separatism among the main minorities.

Yet America does not see itself as just another nation; America is built on a revolutionary past. It was founded on an emphatic rejection of much that seemed ugly in European society; its class-bound social structure, religious intolerance, political oppression, and material poverty. America was to be a refuge for the huddled masses of Europe where they could live as free individuals in social, religious and political liberty, each man with the chance to make his own way in the world. It was more than a nation, it was a venture towards the ideal society.

Of course every nation has a dream-image of itself, but in a revolutionary nation like America or the Soviet Union the ideal is more sharply defined. This has a disadvantage, in that the gap between ideal and reality is more evident, and the national behaviour frequently appears hypocritical. Indeed the ideal will be most strongly asserted when and where it is most threatened, and possibly least observed. Social policy is the field most strongly stamped by a nation's idealized self-image, but it is often an effort to impose this ideal on non-conforming elements (which groups actually dream the 'American Dream' is harder to determine). Hence the aggressively individualist public assistance system in America's South takes no account of the elaborate mutual aid systems among the poor, which make nonsense of personal means tests, and actively persecutes the 'irregular' family structure which it finds.

Yet hypocrisy breeds guilt, and cannot sustain complacency. To most Americans not only is poverty a failure to conform to American ideals, but its persistence challenges the basis of these ideals. It was tolerable, even romantic, as a temporary condition through which particular groups passed as they adjusted to the economic and cultural demands of American capitalism, but 'today's self-perpetuating pauperdom cannot be rationalized' (*Time* magazine, May 1968). The 'rediscovery' of poverty in the 1960s was much more disturbing to the Americans than its parallel discovery in Britain was to the British. The prolific 'poverty literature' reflected America's struggle to accept that poverty exists, and to explain why. Where did the nation go wrong?

The most comfortable way to explain poverty is in the familiar terms of imperfectly assimilated minority groups. But who are today's minorities, now that immigration has been curtailed for over forty years?

Minorities

To begin with, many of the earlier immigrants are still in the population, and immigration did not entirely cease, but indeed revived slightly after the Second World War. Today there are about 9 million people of foreign birth in the USA, mostly Germans and Italians, and 25 million of 'foreign stock', that is each with at least one foreign-born parent. These people, largely Europeans, consider themselves integrated, even if a trifle insecurely, into American society. They have 'made it'. Before the Second World War there were large unassimilated colonies of first generation European migrants, notably Italians, Jews and Irish, mainly in the less attractive districts of the big cities. Today these districts have been handed on to fresh minorities.

One group still arriving from overseas is the Puerto Ricans. Their West Indian home was an American conquest in 1898, and since 1917 the islanders have been American citizens, so they can migrate freely to the mainland. There are only 4 million people in Puerto Rico, compared with one million Puerto Ricans in the USA, two-thirds of them in New York, in a classic linguistic ghetto. Although affluent by Latin American standards, by 'Anglo' standards Puerto Rico is extremely backward, and its migrant population has a formidable task in adjusting to the host society.

A quarter of a million Cubans fled the Castro regime, and the majority are still in Miami. Most of the others are in New York. This migration was received with great sympathy in the USA, and a federal assistance programme was mounted to serve the immigrants, distinctly more generous than Florida's public assistance for its own poor. Apart from the Cubans and Puerto Ricans, the US also has about 5 million Spanish-speaking 'Mexican Americans'. Although some are migrants from Mexico, most are descended from the population living in the South Western states at the time they were annexed from Mexico in 1848. They are a colonial rather than an immigrant minority, protected from assimilation by remoteness and poverty.

We have noted the unifying aspects of America, but inevitably the sub-continent contains large areas and populations, never assimilated by the dominant society, or left behind in its socio-economic development. We are used to thinking of America as the world's most advanced industrial nation, but large parts of it have not yet been penetrated by the Industrial Revolution.

The Industrial Revolution certainly visited West Virginia, but, having ravaged the country for coal, the mines closed and the mining villages were abandoned. Thus it became the first 'island of poverty' charted in the 'rediscovery' of the 1960s.

Other parts of the Appalachian mountains escaped even this intrusion. The natives, descended from some of the earliest English, German and Scots-Irish settlers, still make a poor living as subsistence peasant farmers, or try to 'make it' in the Okie and Tarheel ghettos of metropolitan America. The most disadvantaged Americans are the nation's oldest inhabitants, the half a million Indians, one of whose main distinctions is an infant mortality rate three times the national average; and dominating the minority problem there are of course the 21 million black Americans, first imported as slaves even before the *Mayflower* came, and for centuries deliberately shut out from white society.

About a million Americans are of oriental descent, mostly Chinese and Japanese in the cities of the Pacific coast, the result of migrations from the late nineteenth century onwards. In the past they have suffered discrimination and sometimes vicious persecution, but today, although they remain a distinct group, their income and educational standards are actually above average. The success of the orientals is gratifying to many Americans who use it to show that there is no real colour discrimination. It supports the myth that anybody can 'make it' in America if he tries, as also evidenced by the success of the European migrants. 'My grandfather had nothing when he came. If he made it they [usually the negroes] can.' This rationalized prejudice is known as the Comparative Grandfather theory. Why have the black Americans, and the smaller present-day minorities, failed to make it? The black man's grandfather was a slave, and thus outdoes anybody else's ancestors for disadvantages, but even he was emancipated a century ago.

Until the Second World War the backwardness of black America was part of the general backwardness of the South, i.e. the Southern Atlantic and Gulf Coast states, excluding Texas. Even today half the officially defined poverty in the nation, black and white, is found in this region, which contains about a quarter of the American population. Before the Civil War the region was relatively backward industrially, and the war not only brought massive devastation, but disrupted the plantation agriculture in favour of inefficient peasant holdings. Even the improvement in the social status of the negroes was largely negated by the development of legal segregation. The rise of the New South has been predicted frequently over the last century, and prosperous urban complexes have arisen in North Carolina, and at Birmingham and Atlanta, but on the whole the region was a stubbornly reluctant phoenix. Much of Georgia, Alabama, Mississippi, and Arkansas is still bedded in a mule plough agriculture, where mechanization promises to create only a landless, workless peasantry.

The urbanization of the South was expected to be a second emancipation for the black man but such limited urbanization as occurred was of most benefit to the poor whites. Since the Second World War the booming economy elsewhere has drawn black Americans away from 'Home'. In 1900 they were 90 per cent rural and Southern. Today the majority are urban, and almost half live outside the South. Almost all the blacks outside the south live in cities. mostly big cities. Washington D.C. has a majority of blacks within the city limits (i.e. not including the outer suburbs). Philadelphia soon will have, and New York, Chicago and Los Angeles have large minorities. Continually the black migration reaches farther out, planting minorities in more and more cities hitherto remote from 'the colour problem'.

Within these cities the black migrants occupied the inner zone 'grey areas' replacing the earlier minorities, mainly Italians and Jews. These often continued to own the residential property and operate the small businesses as absentees. Residential segregation progressed as more blacks arrived, until in 1967 the 207 leading cities averaged a segregation index of 86·2 (the percentage of blacks who would need to move 'block'—a quadrangle marked out by the usual grid street lay out—to achieve an even dispersal though the city) (Taenber, 1965).

Most of the black city-dwellers are now young second generation, although first generation recruits still pour in, and the city-born frequently move on to form fresh ghettos in other cities. A black middle class has come into existence, but almost half the black population are still classified as poor. Black standards of living have risen greatly since the war, but for a long time white standards rose faster. In June 1963 the *New York Times* could still report that compared with the average white American, the average black had about half as much chance of staying at school until 18, one third the chance of entering higher education, one third the chance of achieving 'professional' employment, twice the chance of being unemployed, one seventh the chance of earning £4,000 a year, and 7 years shorter expectation of life. In addition the paper could have reported that the black illegitimacy ratio (i.e. the ratio of illegitimate to legitimate births) was 10 times the white ratio, and the black 10 per cent of the American population was credited with half the nation's murders and rapes (most of the victims were also black), two-thirds of the larcenies and three-quarters the car thefts.

Earlier minorities have had a painful initiation into American city life, extending through up to three generations, and the black Americans may fare no worse. However the assimilation of the blacks is currently considered as an exceptionally difficult problem. Daniel P. Moynihan, in his report *The Negro Family: The Case for National Action* talked of the 'tangle of pathology' in black family life, and credited this to the deliberate destruction of the family under slavery. This may well have denied the migrants of a potential strength in adjusting to city life, but present unstable family patterns can be more plausibly related to the present day ghetto environment. This parallels Oscar Lewis's 'culture of poverty', observed in Mexico City and among Puerto Ricans in New York.

It is also argued that the present-day sophisticated industrial economy cannot absorb an influx of unskilled labour, as it did in the past. Yet, despite the fact that the 1960 Census recorded nearly a quarter of the non-white population as 'functionally illiterate' (unable to read a newspaper without difficulty), the black migrants to the cities are better educated than most earlier European immigrants. Probably the main disadvantage compared with earlier immigrant groups, is that black Americans are so visible. Individual blacks are imprisoned for life by the stereotype of their group held by the white majority. Like other stereotypes it will probably fade, but in the meantime it is very difficult to evade. The orientals had almost the same problem, which is forgotten now mainly because they are a much smaller group which has been part of metropolitan America for several generations.

Each immigrant group has passed through a phase of emphasizing its minority identity, before feeling ready to participate on equal terms in the wider society. Black Americans have entered this phase with a particularly fierce intensity, and being such a large minority they have created something similar to the former militant self-consciousness of the European working class. This is a new feature of American life, and has already had a marked impact on social policy.

Summary

Britain is unusual in its homogeneity, not only ethnic and racial, but economic and social. We are very concerned about regional problems, yet there has been no great alteration in the ethnic composition of the population for a thousand years, the country has been almost totally industrialized for over a century, the overwhelming majority of the active population are wage and salary earners, 80 per cent of the population are town dwellers, and few even of the rural population are directly involved in agriculture. Most nations of comparable size have far more diversity. The USA is 60 times as large and 4 times as populous, and has large problems of regional inbalance and internal migration.

The poverty problem of the past tended to be seen as a problem of assimilating overseas immigrants in the big cities. Today the migrants come from the less developed parts of the USA itself, which has awakened America to the existence of these regions. In particular the poverty of the South has exploded across America with the black migration—poor white Southerners migrate too, but they are less visible. The growing self-assertiveness of this large black minority has forced even further attention to this problem, and given it a special flavour.

This is not the entire story of poverty in America, which involves non-migratory non-minority people. Also the migration of disadvantaged minorities is only part of much larger population movements. Three great migrations are in progress: from the agricultural heartland to the industrial periphery (the Atlantic, Gulf, and Pacific coasts and the Great Lakes); from the east coast to the Pacific (California grew from 10 to 15 million population in the 1950s alone); and within each city from the central area to the suburbs. Uprooting and readjustment are a large part of the American experience. The price of this in family breakdown, illegitimacy, suicide, alcoholism, and drug addiction in the growth areas is shared, although unequally, by all groups in society.

2
The comfortable and the deprived

Happy Families

In June 1967 *McCall's* magazine published a lesson in home economics entitled 'How Seven Families Beat the High Cost of Living', by M. T. Bloom. This took 7 American families at the national average household income of $8,000 a year (£3,200) and explained how they managed to make ends meet.

The distribution of incomes in America is far less unequal than in most nations. It is closely comparable to Britain or Sweden, reflecting America's wide spread of education and skill. $8,000 is in the broad zone where the professional-managerial ('white collar') and manual-clerical ('blue collar') classes overlap. 4 of *McCall's* families were blue collar—2 skilled factory workers, a bus driver and a clerk, and the white collars were a doctoral degree student with a working wife, a qualified social worker and a business manager in a low year for sales. If such families can have this income, what can they or America lack?

To a foreigner the striking feature of these families is the abundance of consumer durables. All had cars; 3 had 2 each. An old car was a badge of austerity: 'When I hear a clank in our car—a 1961 Rambler wagon—I get a cold chill . . . it has to last two more years' (student's wife). A dishwasher, two hi-fi radiograms ('semi-needless') and a colour television were also mentioned—although one factory worker decided on puritan scruples not to buy the TV set—but the large refrigerators which supported the families' once-a week shopping patterns, and the telephones which facilitated the telephone advertising of which they complained, were taken for granted.

Four of the families owned their own cars (60 per cent of Americans are owner-occupiers) although they were still re-paying on them, 2 rented houses, and 1 rented a flat. The houses were mostly detached 3-bedroomed dwellings, with rent or repayments about $95 (£40) a month. The American birth-rate traditionally ran much higher than in Western Europe, although it has recently collapsed to near European levels. All the 7 families had 2 or more children and one had 7—that of the Italian-American bus driver from New York, so the houses were well filled.

There was very little budget information about clothes, but from the photographs with the article the families dressed well. While woollens are expensive, cotton clothing and household linens are cheaper than in Europe, and in centrally-heated homes the need for sweaters is slight. The families concentrated most of their 'economies' on food, and boasted of their frugality in the supermarket. However the Italian-American wife accepted her husband's Saturday 'luxuries, such as pop-corn and a gallon of ice-cream for the children'. Food bills ranged from $28 to $50 a week (about £10 to £20).

This is where the wealth of America lies, in a highly efficient manufacturing industry, agriculture and distribution system. The hardware which is the status symbol of the European middle or upper middle classes is the unremarked possession of even the poorest Americans.

The journalist Michael Harrington, while striving to shock middle-class America (Harrington, 1963) admitted that the nation had the world's 'best-dressed poor'. Most food is cheaper than in Europe, although the advantage is mainly in cereals. High protein foods, especially dairy products (except ice-cream) are comparatively expensive. Yet the *McCall's* article was written to demonstrate the financial strains on the average American family and how they were met.

Strains?

Manufacturing industry and agriculture in America may be technologically advanced, generating great production from comparatively little human effort, but the American barber cuts hair in much the same way as a European. Personal service of any kind is delivered in the usual pre-industrial manner. The knowledge taught in American universities may be very advanced, but the teaching methods are substantially the same as Aristotle's. Medical care may save hitherto lost lives, but in the process it becomes more labour intensive than ever. Consequently this is a one-sided affluence, where things are abundant, but services are relatively scarce. In this economy, where manufacturing industry sets a hot pace for wages, the service enterprises have three alternatives. They can go out of business, as many cinemas, theatres, hotels and small shops have done; they can pass on the high wage costs to their customers, giving haircuts, for instance, at $2.50 (one guinea) each, provided that the market will bear it; or they can reorganize to save labour, substituting drive-ins for cinemas, motels for hotels and supermarkets for shops. This last in fact alters the nature of the service, by inserting a large do-it-yourself element. Government and the social services being also service enterprises do not escape this choice, and respond by demanding more tax support, which is only grudgingly forthcoming. This is a large part of J. K. Galbraith's 'private affluence and public squalor' (Galbraith, 1958), for the cost of public services tends to rise much faster than that of private consumption without any improvement in the services provided.

Services are the area where the American family is least at an advantage compared with Europe. The seven *McCall's* families seldom went to a cinema, or ate at a restaurant, and only went away on holiday when they could stay with relatives. Higher education for parents or children presented a severe economic problem, not only of maintenance but of tuition fees. However all the families showed a great eagerness for education or training, and met these costs even if at a sacrifice, for these were model families, chosen to exhibit the Puritan ethic.

The great area of unmet need was medical care. Some of the families enjoyed free or subsidized health insurance from the husband's employment, but for instance:

'Jennifer [the social worker's daughter] needed an expensive stop-gap operation for a congenital heart condition; she will need another major operation when she is ten. Even with Blue Cross and Blue Shield [medical insurance] they had large out-of-pocket expenses and had to put off necessary dental work.'

'In 1965 Ray [an aircraft worker] was laid off and soon after went to the hospital for a month with San Joaquin valley fever. The sickness and its sequel cost $6,000. Fortunately his previous employer's health plan paid $5,000; but the $1,000 was still a great dent in their savings.'

'Unfortunately [the employer's group medical insurance] does not cover dental care and one of the [bus driver's] girls needs about $500 worth of orthodontia.'

This is simple budgetary reporting. The article was not concerned with promoting publicly financed medical care. However it could not help illustrating the lack of 'social incomes' in the USA. Money incomes must cover needs usually met in whole or part from public social services in Europe, particularly medical care and low-income housing. Hence in America shortage of money has far stronger adverse implications for health and shelter.

Poor Americans

If $8,000 is the average family income, what is the poverty line? J. K. Galbraith writing in 1958 used a cut-off of $1,000 per annum, which yielded 10 per cent of the population in poverty. However, this was below any state public assistance budget for minimum household needs. In 1966 the basic standard for a family of 4 on public assistance ranged from $1,326 in Arkansas (the total grant was well below this, since only a proportion of officially recognized need was met) to $2,213 in Washington D.C. On the basis of each of the various state assistance budgets, 17 per cent of all households were substandard in 1959 (Miller, April 1963). About 4 per cent of the population actually received assistance.

But state public assistance is noted for parsimony. The 'official' poverty lines usually used are those suggested by federal agencies. In 1964 the President's Council of Economic Advisers set $3,000 for households of two or more as the poverty line, yielding 9·3 million poverty-stricken families; that is 30 million Americans, or 20 per cent of the population. This estimate has since been refined by Mollie Orshansky, a government statistician, to allow for family size, and urban-rural differences. A rural family was estimated to need 40 per cent less money income than its urban counterpart (Orshansky, 1965). Orshansky's standard ranged from $880 for an unattached woman over 65 living on a farm to $5,100 for an urban family of 7 or more headed by a man under 65. The urban family of 4 was reckoned to need $3,130. Although this revision changed the composition of the poverty group, its size was much as the Council of Economic Advisers estimated. In fact it embraced more people in fewer families (43 per cent of Orshansky's poor were children), with 35 million poor.

These federal poverty lines imply a concept of 'absolute poverty', working from Department of Agriculture economy food plans—about 2 shillings per person per meal. In Orshansky's definition, one is poor when the minimum nutritionally adequate diet costs more than a third of one's income. This is a static concept, whereas the usual understanding of poverty, linked with what society sees as the minimum 'decent' standard of living, is relative and dynamic. To set one's sights on a static poverty line may well create an impression of success without in the long run relieving relative deprivation.

Who are these poor, and what is their living like? This depends on one's choice of poverty line. In European nations the income maintenance system usually prevents many people falling very far below the official established poverty line. However, in America the system is not nearly so comprehensive. Only a quarter of Orshansky's poor received public assistance, principally because her standard was far above that set by most state public assistance departments. Also eligibility for assistance is limited by many non-financial factors, principally employment, family and medical circumstances. Finally, the poor are not encouraged and are often not eager to seek assistance on the terms on which it is offered. Of the 9 million families with annual incomes under $3,000, 3 million had from $1,000 to $2,000, and 2 million had less than $1,000 (just over £400).

With a $3,000 poverty line about a quarter of the poor are 'non-white', that is, mostly blacks and a few American Indians. This represents about 40 per cent of the non-whites in the nation. But, as one lowers the cut-off, the non-white proportion rises sharply. In some counties in Mississippi the *average* negro income is under $1,000, and this is also true of many Indian tribes. The Mexican-Americans too figure increasingly at the lowest incomes. Migrant farm workers, the poorest of whom tend to be Spanish speaking or black, had average earnings in 1962 from all sources of $1,123. Leaving out the 'Anglos', the figure would be well under $1,000.

The above examples of very low incomes are all rural. Despite the shocking city slums, poverty is concentrated more heavily in small towns and rural areas. (30 per cent of all families, and 46 per cent of families under $3,000 are rural.) Most of the rural poverty, white as well as black, can be traced to the South.

The Council of Economic Advisers counted one third of the poor as old people; Orshansky reckoned a fifth. In any case, this is the group best served by the income maintenance system, most of them drawing benefits from the federal old age insurance scheme. Thus the aged occur less among the lowest income groups, and do so usually only where they have other poverty characteristics such as black skins or rural homes. On the other hand children, prominent among the Advisers' population, take almost half of Orshansky's poor. To her, poverty, particularly in cities, is mostly a combination of low earnings and too many children. Children figure increasingly as one goes down the income scale, and absorb most of America's worst deprivation. The extremes are usually in large families in broken homes, that is, when the father has been lost by death, desertion, or divorce, or where there has never been a father. In 1965 the income of fatherless families with 5 or more children *averaged* 41 per cent of Orshansky's poverty standard. In Florida such a family would qualify for $85 (£35) a month *maximum* public assistance grant.

In 1963 the Department of Labor analysed the employment status of families with under $3,000. 30 per cent were headed by persons completely outside the labour market; most of them old people or mothers with young children. 16 per cent of household heads had experienced unemployment. 14 per cent had passed in and out of the labour force without ever being classed as unemployed. 30 per cent had been in employment all year. Unemployment is clearly an important cause of poverty, but increased economic activity alone is not nearly enough to vanquish it.

Galbraith distinguished 'insular' and 'case' poverty. He portrayed certain 'islands' of communal poverty in remote areas such as West Virginia, and special 'cases' of inadequacy in otherwise affluent environments. Since then (1958) we have come to see the poverty 'islands' as extensive backward regions, such as most of the rural South, where poverty is normal and traditional as in Southern Italy or the Balkans. In rich mainstream America, poverty indeed attaches to special circumstances, like old age, ill health, unemployment, broken homes and extra-large families, but these circumstances are so common in some city districts as to indicate poverty-prone environments. Most of these high-risk areas are in fact peopled by first- or second-generation migrants from the poverty regions, so that insular and case poverty are strongly interlinked, and closely bound up with America's cultural, ethnic and racial minorities.

Finally, what is it like to be poor, American-style? The living pattern of the $8,000 families gives us a clue. The abundant durables are still there among the poor, but second-hand and less reliable. In Tunica county, Mississippi, one of the poorest in the nation with nearly 80 per cent of the families under $3,000 in 1967, 37 per cent of the families had washing machines, 48 per cent had cars and 52 per cent owned television sets. Car ownership is so universal that it has driven out public transport in many areas, such as Los Angeles, where even the public assistance department makes grants for car purchase so that its recipients can find work. Petrol ranges from 2 to 3 shillings a US gallon.

The good standard of clothing among the poor is credited with making them 'invisible' (Harrington, 1962) in the 1950s. This is true of the city poor, who wear the sneakers, jeans and tee shirts that are America's leisure uniform, but in the deep South poor children seldom wear shoes except for school and church.

The various federal poverty lines by definition provide a nutritionally adequate diet. Below this standard high protein foods become more difficult to obtain. Dried milk replaces liquid, and chicken becomes the main form of meat. One mother described to me her experience on public assistance as 'six months on chicken backs'. Butter and cheese each at 6 shillings a pound leave the diet, but margarine and eggs are plentiful enough for all except the largest city families. Harrington felt confident enough to report 'to be sure, the other America is not impoverished in the same sense as those poor nations where millions cling to hunger as a defence against starvation. This country has escaped such extremes.'

Having myself lived on the edge of a very poor rural district in North Florida, I am not so sure. In 1968 the Citizens Crusade Against Poverty reported in a well-documented study *Hunger, USA* that 10 million Americans were chronically undernourished. Most of these lived in the South, although the Indians of the desert states were perhaps the worst starved. A team of doctors from the Field Foundation in 1967 met Mississippi children 'who don't get to drink milk, don't get to eat fruit, green vegetables or meat. They live on starches—grits [a type of semolina], bread, Kool Aid [a soft drink from sugar and fruit flavouring].' Deficiency diseases were common.

Housing conditions vary dramatically. The slum districts of the West Coast cities are spacious, pleasant-looking estates of detached houses, simply older and perhaps less well maintained than other housing. In the South whole villages live in a picturesque squalor of unbelievably dilapidated shacks. Philadelphia has a Manchester-style grim acreage of red brick Victorian slums, grouped around narrow courts full of flies and refuse. Especially in the cities where the migrant pressure has itself built up a housing shortage, housing is the main visible distinction between rich and poor. Public housing (i.e. council houses) is comparatively rare.

What the poor lack almost entirely are services, except those provided by friends, family, charity and public assistance. Schooling is free from the age of six, but nursery schools are mainly for the fee-paying middle class, who usually start their children's education at the age of four. Higher education can be 'worked through', particularly since courses seldom demand continuous full-time study, but this involves the poor in great sacrifice, persistence and stamina. Medical care is usually available free for medical catastrophes, either from the doctor's charity or from public assistance, but 'minor' care such as dental care or spectacles for short sight is difficult to obtain. In 1965 a medical examination of low-income nursery school pupils in Jacksonville, Florida, reported that of 1,005 children checked, 52 per cent were anaemic, 42 per cent needed dental care, 32 per cent had hearing defects, 25 per cent had eye trouble and 5 per cent were partially blind (*Weekly Compilation of Presidential Documents*, August 1965). This lack of preventive care largely accounts for huge differences in infant mortality between income and racial groups, the black rate running twice as high as the white, which itself is higher than most West European rates. However certain high-income American counties record some of the lowest infant mortality rates in the world.

In summary, America's poor range from income levels which are modest by European standards but scarcely constitute hardship, to levels where even European basic necessities are scarce. The life-style of the poor is a pale reflection of the average household, still comparatively well stocked for durables, adequately clothed, with hunger only at the lowest levels, although seldom with the best balanced diet, and with very limited command over services, except where publicly provided. The condition of poor people is most disturbing to Europeans in those areas of life where European nations have intervened most extensively in the market system, health and housing. This is where America's poor are most vulnerable.

3
Divide and rule

The federal role

In the beginning, there were thirteen rebel colonies, and they created a military alliance and then a federal government. This was to win their independence and preserve it, not to subordinate themselves to another master. Accordingly, the powers allowed to the federal government were strictly limited by a written constitution, which, though now nearly two hundred years old, is substantially unaltered. Further powers were named for the member states, which also held all residual powers, except in those areas of personal freedom which the constitution expressly protected from all government (these last constitute the so-called Bill of Rights). The federal legislature, Congress, could of course alter the constitution by special procedure, but this was not easy. America, then, is governed largely according to a set of written rules framed to constrict authority, especially federal authority. In spite of this, a major theme in the history of the country has been the tightening of the federal union, by civil war, taxes, and legal action, pulling the loose association into something like a co-ordinated nation.

In principle, the present-day federal government looks after foreign relations, defence, the national debt, natural resources (including farm subsidies), the Post Office, space research, ex-servicemen, and water and air transport, while the present fifty states are sovereign in education, highways, health services, public assistance, police, sanitation, fire services, parks and recreation, and the penal system. In practice, the functions of the two levels of government elaborately interpenetrate, more a marble cake than a layer cake.

In the last resort the federal government can impose its policies by force, constitutional niceties notwithstanding. The retention of the southern states in the union stems from their century-old defeat in the Civil War, and several of them might possibly opt for secession now if they could. As it is, the federal government has sufficient support in the nation at large to try to bring the social and political structure of the South more into line with generally accepted ideals. Naturally the southern states are in varying degrees unco-operative. One southern governor recently numbered the federal government as one of the great enemies of freedom, along with 'International Communism, the United Nations and racial agitators'.

However, open resort to force cannot be an everyday feature of federal-state relations, though it was used as recently as 1957 at Little Rock, Arkansas. Legal pressure can be brought to bear through the interpretation of the constitution in test cases by the US Supreme Court. The Court is appointed by, but acts independently of the federal government, and one of its main functions is to see that the constitution is observed—it is the referee between the federal and state governments and the individual citizens. Lately the referee's decisions have tended to run against the state governments, usually in favour of citizens' rights, but also strongly consistent with federal social policy. The most celebrated Supreme Court decision was probably the 1954 school desegregation ruling, although it is still not fully enforced. Possibly the most extreme intervention in state affairs was in 1966 when the Court ordered the re-election of the Florida state legislature on a new constituency map—the first time that an outside authority had dissolved a state legislature since the Westminster Parliament passed the Massachusetts Act in 1775. The legal lever against the states in all the major cases has been the defence of constitutionally protected individual freedom. This is capable of considerable extension within the wording of the Constitution whereas federal powers are closely defined.

Yet Court intervention can still only be intermittent. For day-to-day supremacy the federal government relies on its financial power. It commands the most effective tax collecting system in the nation, the Internal Revenue Service, which raises most personal and corporate direct taxes. The states can also raise what taxes they please. Some have state income taxes, but most rely on relatively inelastic sales taxes (a regressive form of purchase taxes). Furthermore, the least enthusiastic supporters of federal policy are often the poorest states, least able to spurn subsidies. The grant-in-aid has been a prominent feature of federal policy since 1935, whereby the federal government virtually bribes the states into operating various social programmes. The grants are usually on strict conditions, and if a state wishes to keep its freedom it does not get the cash. Often this develops into a subtle game, where the state tries to get the maximum subsidy consistent with the least change in policy, such as by reclassifying existing welfare recipients to get the higher grant attached to a new welfare programme.

Before the 1930s the federal government was not very

interested in social policy. The great enlargement of the power of the federal government in the last thirty-five years has been part of the general enlargement of all government activity, prompted by the Depression and the Second World War. This has brought the federal government into fields hitherto rather sparsely cultivated by state and local government, and generated a close interweaving of federal, state and local activity. The federal dominance based on the grant-in-aid, has led to such national cohesion as now exists in public assistance, public health, and similar services. Until the 1960s the states raised no serious objections—indeed they welcomed the money, particularly in the South. It was not until Washington began to respond positively to the Civil Rights movement that a serious states rights' backlash began. The federal government has responded by threatening to withhold grants, and being ready to use force in support of an astonishingly liberal Supreme Court. It has also been driven to circumvent the states, and to bring influence to bear directly on local governments and communities. The War on Poverty was one of the first exercises in this.

Federal government

The architects of the federal government were reacting against the government of George III, yet paradoxically created a system that remains more like an eighteenth-century constitutional monarchy than anything Britain has today.

In 1776 the British 'constitution' was in transition, with Parliament having won a fragile legislative independence, but with the monarch still in control of executive power. This 'separation of powers' was built straight into the American constitution, with Congress in the role of Parliament and the President as an elected monarch holding office for four years at a time. Like an eighteenth-century king, he appoints and directs the executive officers of government, chooses his own 'cabinet' of personal advisers, and can veto Acts of the legislature.

Even the separation of Parliament into two chambers, reflecting the division of the English social structure, was copied in the separation of Congress into Senate and the House of Representatives. However, the Americans made both chambers elective, which, with an elected President, created an enduring balance, each part with some democratic sanction. In contrast, the growth of democracy in Britain forced a concentration of power on the House of Commons. As a final element, America follows the English concept of an independent judiciary, and uses it, as we have seen, as a constitutional watchdog.

This fragmentation of authority is obviously cumbersome and stifles strong action except in overriding crises, but at least it provides something acceptable to the states. Also it has been flexible enough to survive longer than most systems of government, with very little formal amendment.

Federal politics

Early in its history, the United States developed a two-party system misleadingly similar to England. However, within the context of the separation of powers, it is obviously useful, but not essential, for the President to belong to the majority party in Congress, and party discipline is not so important, for the executive cannot be overthrown by an adverse vote. None-the-less, a strongly organized party system would offset the division of formal authority, and promote more effective government. The formal division only matters because there is no strong party system outside election time.

The Democratic party is broadly to the left, advocating stronger government and more adventurous domestic reforms and foreign policy—an unfortunate combination in some ways, since Roosevelt, Truman and Johnson each ultimately had to sacrifice his social policies for the sake of a war. The combination is, however, quite logical, for the wars have been fought in pursuit of much the same ideals as sought by internal reform.

It is somewhat surprising to find this 'left-of-centre' grouping the dominant party in the South, which for the last century has been almost a one-party region. The 'Dixie-crats' favour Roosevelt-style federal aid to poorer regions, but are well to the right of Johnson-style race relations. The Democrats are in fact a loose coalition of Southern racialists, racial and ethnic minorities in the northern cities, and organized labour—different things to different people in different places. Very confusing—but then does the Conservative and Unionist party stand for the same thing in London and Londonderry? In so large a nation, regional issues frequently cut across party concerns, and it is remarkable that some semblance of a two-party system does not operate from coast to coast, instead of a multiplicity of regional interest groups. However, the Democratic party in particular has frequently been paralysed by its internal conflicts.

The Republican party is the more conservative; traditionally opposing strong and expensive government, supported by big business and the more prosperous agricultural interests. Unlike the English situation the conservatives are the natural minority, controlling only 2 out of the 12 Congresses from 1945 to 1968, and with a majority of the state governors only 1947-8 and 1951-4. However,

American voters justifiably tend to follow personalities as much as parties, with many 'split tickets' at elections. For instance, a person may vote Democrat for his Congressman, and Republican for President. Consequently, from 1956 to 1960, and again in 1969, the majority in both chambers of Congress was Democrat but with a Republican president.

It can be argued that this looseness of party ties, and the prevalence of personal factions is not inherent, but exists because there have been no real party issues in post-war America. Roosevelt won the 'New Deal' in the 1930s, and in the post-war era both parties were broadly content to let the federal government's role in social policy lie as he had left it. From 1945 to 1964 the 'politics of consensus' were virtually unchallenged. Goldwater broke the pact in 1964, only to bring the Republicans to their worst electoral defeat in a century. By 1968 both major candidates were once again matching personalities more than policies. Yet the events of the 1960s have restated the Democrats' claim to be a reformist party, have strengthened their support in the big cities, and gathered almost all the black vote, while simultaneously driving away many Dixiecrats into a regional third party. The Democratic coalition has been redrafted, although at the cost of losing office to a quietist Republican Administration. In 1968 there was a more coherent division of electoral support than has been visible for a generation. However, in the American situation this will always be difficult to sustain.

One might imagine that the most progressive institutions in American government in the 1960s would have been those with the closest direct contact with the people, such as the state governments or the House of Representatives (the Senate is much smaller, and the senators are elected at large from their states, not from individual constituencies). In fact, these have been the most conservative bodies, both over-represented with agricultural interests. Also the 'House' is dominated by its senior members as committee chairmen, usually long-serving Dixiecrats from one-party states. The Senate tends to be more progressive, perhaps because, being elected state-wide, the senators are more responsive to urban interests. Only once has Congress as a whole been ahead of the President—in the 1964 Civil Rights Act. Usually the progressive initiative has come from the President, prodding the legislative mule. The greatest surprise has been the progressive zeal of the Supreme Court, largely the work of Chief Justice Earl Warren, who set a pace that left even the Civil Rights organizations breathless.

States and localities

The state governments are all, except Nebraska, small replicas of the federal government, with a Governor as executive head, a two-chamber legislature, a Supreme Court and a written Constitution. There is no standard local-government structure. However, the broad principle is English-style local autonomy, with most of the powers allotted to the states by the Constitution, delegated to local government under state supervision. Most of the states are divided into numerous counties, and most of the population centres of any significance are incorporated 'cities'. There are also frequent 'special districts', single-purpose local bodies outside the general multi-purpose structure. The commonest of these are the school districts, operating the primary and secondary school system, with their own taxing powers, like England's School Boards before 1902.

Probably the most striking difference in city government from Britain is that the Mayor is not a ceremonial figure-head, but usually the major political personality in his community. There is no British equivalent to Mayor Daley of Chicago.

4

The pin-table welfare state

American welfare has been compared to a pin-table. The prospective client is shot like a pin-ball into an arena where several different pockets offer him a refuge, some far more valued than others. He rolls about hoping to be caught up, slowly dropping down among the less eligible pockets, until perhaps he drops right through the system without finding support. Sometimes he will bounce, that is be referred, from one pocket to another, and he may keep bouncing for some time.

This model is true to some degree of every country's welfare system. Each tends to evolve from services developed for special groups, such as ex-servicemen, school-children, old people, or the very poor. In Europe, especially Britain, these have now been fused into fairly comprehensive systems of specialized services, with more emphasis on the type of service than the identity of the recipients. In America the system is still based more on client groupings. For instance, health services are fragmented among many different agencies, which include them in integrated parcels of service, each for a particular clientele. Some groups are well favoured, others less so, and some find it difficult to locate any help. The French welfare system is at an interesting intermediate stage, struggling to escape from its tenacious sectional origins.

American welfare (using welfare in the broader European sense and not, as in America, as synonymous with public assistance) also refutes the maxim that a nation has the social services it can afford, rather than those it needs. In Europe the richest nations tend to spend the highest proportions of their wealth on social services. In 1963 the USA, 30 per cent richer than the richest European nation (Sweden), spent 7·6 per cent of its national consumption on social welfare, excluding education and public housing. Britain spent 13·5, Sweden 17·5, France 18·6 and West Germany 21·0 (*The Cost of Social Security*, ILO 1967). America's proportionate expenditure has risen from under 6 to almost 8½ per cent over the decade 1957-67, raising its performance relative to most of Europe. However, it seems unlikely to catch up in this century.

Public assistance

This is the bottom range of pockets on the pin-table. It is not, and does not pretend to be, a safety net.

The system is directly descended from the English Elizabethan Poor Law, and retains some features now vanished in England. Assistance is 'parochial' in the sense that there is no single national system, but a series of separate state, and to some extent county systems. In 1968 all the states had residence conditions for assistance grants, and indigent non-residents were often returned at public expense to their place of last settlement.

American public assistance departments also offer a comprehensive range of services, as did the Poor Law, including cash payments, medical aid and institutional and foster care for children, and more recently free food, home helps, day nurseries, employment training, family counselling, consumer education, and family planning advice. Finally, public assistance in America has maintained a deterrent attitude to its clients until very recently. It was particularly noted for forms of harassment such as 'midnight raids'—surprise inspections of fatherless families to check if there were men in the homes.

These Elizabethan survivals may not persist much longer. Several test cases have found residence laws unconstitutional, and by 1970 almost certainly none will exist in the USA. The cash and services offered by public assistance departments are beginning to be separately organized and delivered, and employment training has already passed to the general state employment services. The first signs of a break-up into universally available specialized services are beginning to show. Also, there is a current effort to reverse the deterrent image of assistance, and the grosser forms of harassment have been outlawed.

Until the late nineteenth century, assistance was handled entirely by the counties. State governments became involved from 1867 onwards, largely with institutional care for the blind, the handicapped and children. After 1911, certain programmes of cash assistance were developed in many states which separated the blind, aged and fatherless families for more generous treatment than other paupers. Unhappily this 'categorical welfare' drew a persisting invidious and arbitrary distinction between the deserving and undeserving poor.

Federal intervention through the grant-in-aid dates from the 1935 Social Security Act, which, with its many subsequent amendments, still governs most US income maintenance services. Federal grants were provided for each of the three existing 'categorical programs', which became known as Aid to the Blind (AB), Old Age Assistance (OAA) and Aid to Dependent Children (ADC). This last was renamed Aid to Families with Dependent Children (AFDC) in 1962. Other financial assistance, termed General Assistance (GA) attracted no federal subsidy. The main conditions imposed on a federal grant were that the money had to be spent on a carefully defined population (and defining a fatherless family is very complex), the programme had to operate throughout the state, and the state or localities had to 'match' the federal money on a complicated formula. Even the provision of these comparatively small matching funds inhibited some states, but by 1958 all were operating these three programmes, plus a later small programme usually called Aid to the Permanently and Totally Disabled (APTD). Examples of the rates of benefit to AFDC clients are contained in Chapter 2.

Federal intervention paradoxically enhanced state power, for the money went to the states, not to counties, and the state governments were responsible for enforcing the conditions attached to the grants. Gradually two systems emerged, *state administered* and *state supervised*. A state administered system operates all assistance through the state civil service, with no local autonomy. State supervised systems vary from minimal state intervention to vestigial local control (usually local approval of personnel appointments). Frequently the state will run the categorical programmes while the counties look after general assistance, which means that in many areas (e.g. most of Florida) there is no general assistance.

Federal aid has been extended considerably in the last decade, to pay half the administrative cost of categorical programmes, and 85 to 95 per cent of the cost of social work services to AFDC families. The definition of fatherlessness had also been liberalized, but not all states have taken advantage of this. Medical Aid for the Aged (MAA) in 1960 introduced the concept of the 'medically indigent'; persons unable to meet their medical bills but not otherwise eligible for assistance. The concept was extended to all age groups in the 1965 Medicaid plan.

In the not-so-distant future probably all assistance will be 100 per cent subsidized by the federal government, with standarized regional benefit scales. The present confusion of categories, and the cruel distinctions it makes between cases in quite similar circumstances, is probably a transitional phase, such as existed in England from 1934 to 1948 while the central government gradually drew responsibility for assistance from local authorities. However, in America this is a slow and halting transition.

The public assistance systems of Western Europe are also usually local and comprehensive (i.e. combining cash and services) and often categorical. However, the great development of 'universal' income maintenance programmes, through social insurance and family allowances, has reduced European (but not British) public assistance to relative insignificance. Britain's limited universal services force it into a dependence similar to America on means-tested assistance schemes, but uniquely Britain has fashioned a national, specialized, non-categorical (and since 1966 partially disguised) assistance system.

Social insurance

Social insurance is a device to make the gift of public money to individuals socially acceptable. It avoids the personal indignities of the public assistance means test, by presuming need among defined groups of the population, such as the sick or the elderly, and granting them benefits on the pretence that they have paid for them. The pretence is based on simulated insurance contributions. The façade of insurance is more strongly maintained in America than in most nations, but even here the resemblance to commercial insurance is weak.

The device came to America in 1935 50 years after its first use in Germany and nearly 25 years after its introduction to Britain. However the first benefits were not drawn until 1940, and initially only 60 per cent of the population was covered. The era of 'Social Security', which means social insurance in the USA, is largely coincident with Britain's experience as a 'Welfare State'. It is a post-war system, in sharp contrast to the system still remembered by the older generations.

There are two main American schemes, a state system of short-term unemployment compensation, and a federal old age, widowhood and disability plan. Such federal activity was probably unconstitutional in 1935, but during the Depression the Supreme Court was prepared to be broad-minded. Thus, although benefit rates are good, by British if not Continental standards, the risks covered are few. It is far from 'cradle to the grave' insurance, but, rather, an extra pocket in the pin-table for the more fortunate elderly.

The federal scheme, Old Age, Survivors, Disability and Health Insurance (OASDHI), is basically an income and medical benefit arrangement for the elderly, with incidental cash benefits for widows, orphans, and long-term disabled. There are no general sickness or medical insurances, no maternity or death grants, and no family allowances, except in the form of income tax exemptions. Insurance benefits and contributions are both wage-related, the contributions being partly an employer's payroll tax and partly a rudimentary personal income tax. This permits a considerable redistribution of income, by adjusting the benefit formula to favour the lower rate contributors.

The minimum benefit (1968) is $55 (£23) a month for a single person and the average benefit for a retired couple is $159 (£66) a month, about 60 per cent of the federal minimum wage for a full-time worker. This is approximately the federal poverty line for an elderly couple, but well above most public assistance limits.

American pensions are not 'dynamic', except in so far as wage-related benefits have a built-in relation to living costs and standards. Also there is an erratic dynamism in that Congress revises benefits at about five-year intervals.

The growth of social insurance still leaves a considerable dependence on public assistance, as in Britain. But, whereas most British means-tested allowances supplement low-insurance pensions, in America social insurance alone caters for nearly all the elderly, leaving assistance mainly to families with children. Furthermore, the original target group for the ADC scheme, respectable widowed mothers, has also been taken up by insurance. A less 'deserving' group of divorcees, deserted wives and unmarried mothers has inherited a scheme once designed to give some slight privilege to the more worthy poor.

Other pockets on the pin-table

Possibly the most coveted pocket is the capacious Veterans' Administration. Even before the Vietnam War (1963) there were 22 million ex-servicemen in the nation, with 60 million dependants, 44 per cent of the total population. The VA, the nation's largest welfare agency, runs 169 hospitals and 18 welfare homes, primarily to treat service-connected disabilities, but in fact meeting almost all ex-servicemen's hospital needs. It has been a pioneer in psychiatric care. In addition to medical care, it provides an assistance scheme for indigent ex-servicemen, educational grants, insurance schemes, loans and mortgage facilities, burial grants, and an American flag for each veteran's coffin. This importance of military-related welfare is typical of a nation with a relatively limited general welfare system. Ex-servicemen can establish a claim on the national conscience long before a wider sense of mutual responsibility takes hold.

The federal Public Health Service provides free medical care for American Indians, merchant seamen, federal prisoners, Presidents of the USA, and certain others. State Health Departments, under federal subsidy, operate various environmental and personal preventive services such as infant welfare clinics. The states also provide mental hospitals, while local communities are beginning to operate mental health clinics. Otherwise hospitals are private non-profit institutions, charging an economic fee to their patients. Doctors are usually private entrepreneurs, and are seldom salaried by hospitals, i.e. the same doctors serve in the hospital and the community.

Public housing accommodates less than one million families (2 per cent of the population) among the lowest income levels. However, higher up the income scale the Federal Housing Administration has helped about 9 million families to become owner-occupiers, since 1937, through mortgage insurance, and issued 28 million house improvement loans. Several schemes for low-interest loans for house purchase, and rent subsidies in non-profit housing, have been added recently. Various federally subsidized slum clearance programmes have operated since 1949, although not always to the advantage of the poor.

In almost every American town the numerous private welfare organizations extend an effective system of voluntary taxation over most of the population, through joint fund-raising drives. These mostly operate through workplace representatives, who enjoin their colleagues to pledge a fixed proportion of their earnings to the 'United Good Neighbors'. Voluntary and statutory organizations also co-ordinate through Health and Welfare Councils which exist in most big cities. Possibly 20 per cent of American health and welfare expenditure excluding social insurances comes from private and voluntary sources, most of it going to social work services, principally recreation, family counselling and child care. Much of this is consumed by the middle class, and little falls to less stable low-income groups. However, voluntary welfare fosters a high degree of interest and participation among some sections of the public, and creates a web of formal and informal contacts which was the basis of the first federally sponsored community action enterprises.

About 15 per cent of retired Americans receive pensions from former employers, and about 30 per cent of present workers are covered for future occupational pensions. The most important occupational schemes are group medical insurances, which help provide limited hospital coverage for most Americans, and widespread physicians' insurance.

Income tax is an even more important method of income adjustment than in Britain. As well as wife, children and mortgage, the American taxpayer can charge exceptional medical expenses, and most charitable contributions against tax, representing a considerable indirect subsidy.

One of the most effective welfare measures is the federal minimum wage ($1·60 an hour), sometimes supplemented by higher state minima. About half the non-agricultural work force is covered. This is a massive public intervention in the free market, yet strangely America finds it consistent with its notions of free enterprise. Most recent anti-poverty measures have concentrated on reinforcing the wage-price system, even to fixing the price of labour, rather than providing social incomes unrelated to employment or extensive free or subsidized public services.

Education

No outline of American social services can be complete without mentioning education, although Americans seldom classify this as a social service. The European nations inherited a nineteenth-century structure with a prolonged narrow academic education for the children of the rich, and a shorter, more practical instruction for those of the not-so-rich. Today this survives as separate systems for the clever and the not-so-clever. America never shared this tradition. Universal free public education existed in most settled states by 1830, deliberately shaped to integrate new settlers into a common culture, irrespective of class, creed, national origin or sex. This rejection of European elitism distinguishes America from all but one European state, itself a revolutionary nation, the Soviet Union.

Racially segregated schools still survive in the South, but this is usually deplored throughout the rest of the nation, as are the black ghetto schools which many municipalities try to diversify with elaborate cross-town exchanges of children by bus. Private schools exist on the east coast to a small extent, and the Irish immigrants brought church schools, which are still viewed with some suspicion. Otherwise education is wholly comprehensive, co-educational and secular. Also, in response to the traditions of frontier society, American schools teach useful skills to all pupils, such as car driving (which virtually eliminates private driving schools) and typing (which makes marking essays much easier). There is emphasis on acquiring social skills, and a wide range of studies is carried throughout the school career, avoiding the intense specialization of the English Sixth Form. This breadth of curriculum is enhanced by the absence of any state or national school-leaving examinations.

Compulsory schooling usually runs from the age of 6 to 16, but nearly 70 per cent of pupils stay at school to 18. Early leavers are known even in official literature as 'drop-outs'. 40 per cent of school-leavers enter higher education. In many wealthier suburban communities, the entire school population goes on to university. About 20 per cent of the age group get B.A. degrees, not all as good as English B.A.s but better than most English non-graduate qualifications.

However, higher education is class-selective, for not only are student maintenance grants rare, but most institutions charge high tuition fees. The public universities recover only about half the cost of instruction from fees, but the private universities charge most of the cost to their students. Unlike primary and secondary education, the private sector in higher education is large (45 per cent of students), including many sectarian colleges.

Most students manage financially only because of the extensive opportunities for part-time work, even in term time. The degree structure also permits interruption of courses and part-time study. Much student employment is created by the universities themselves, which reserve campus jobs for students, as refectory staff, caretakers, library clerks, secretaries, film projectionists, research assistants and so forth. This form of educational maintenance is the model for many anti-poverty programmes.

The main shortcomings of the very progressive elementary and secondary system are the variability of standards, and the poor educational relationship with many 'low-income' pupils. The first stems from the tiny administrative units on which the system is based and the residential segregation of the big cities, which defeats much of the integrative effect of comprehensive schooling. The second is a problem which Europe has side-stepped by never attempting to give a meaningful education to the majority of its older children.

Reluctant welfare

A European cannot but wonder that so rich a nation has such rudimentary social services, apart from education.

Much has been said of the lack of class consciousness in America, possibly due to social mobility, and the geographic, religious, ethnic, and racial fragmentation of the population. This inhibited the growth of self-help organizations, such as the English Friendly Societies or the French *caisses* on which the contemporary European social security system was built.

However, we cannot claim that this is an unusual Anglo-Saxon situation. Britain, when it was the world's richest nation sixty years ago was equally backward in welfare services in relation to much of the Continent, and still has a fairly exiguous welfare system by Continental standards. America inherited Britain's puritan self-righteous, laissez-faire economics, and distrust of all government, intensified by the experience of the Revolution and the western frontier. Both the British and the Americans adapted grudgingly to the more collectivist needs of industrial society, and so long as they experienced rapid rates of economic growth, they were happy to postpone their problems to a bright future where everything would solve itself. In the industrial sense Britain ceased to be a 'frontier society' about 1870, America about 1920.

Britain began to organize itself as a mature industrial nation with the advent of social insurance in 1911, America in 1935. Both have had problems in reconciling freedom and welfare, but Britain has been pushed further into recognizing social welfare as a form of positive freedom.

At present even senior welfare officials in America are often baffled by contact with European 'universal' welfare. The head of a big city public assistance district once welcomed me to his office with a story of his Air Force days in England after the war.

'We had a baby, and they tried to give us some free milk. Of course, I didn't take it, as I had a captain's pay and we were quite all right. But it was very, very nice of them to offer it to us.

'Then they tried to give us more money, when the baby was born. Of course I didn't take it, but it was very nice of them to think of us like that.

'I don't know why they did this. Did you?'

5
The decade of opportunity

Poverty rediscovered

It is common to label the various decades of the twentieth century by their dominant characteristics, as though the calendar itself divided one phase of human history from the next; the gay twenties, the romantic thirties, the heroic and austere forties. The 1950s will probably be remembered as the decade of affluence, perhaps of complacency.

It was an enviable decade in many ways. The great disasters and upheavals of the previous two decades were over, at least in the developed nations. The Cold War grew less chilling, the economy was stable and expanding, the social reforms put forward in the thirties and forties became accepted realities in both America and Europe, non-reactionary conservatives held power in most western governments, almost every American and British home obtained a television set, and poverty was believed to be a vanishing anachronism. A British election was won on the slogan, 'You never had it so good'. It could have served the Eisenhower administration just as well.

Yet the very name widely chosen to describe this period was the title of the book that first seriously challenged its complacency, J. K. Galbraith's *The Affluent Society* (1958). This book talked about poverty, although it was not the main topic. It gently asserted that 10 per cent of the American population was deprived. By 1962 Michael Harrington had written a polemic, *The Other America*, listing various poverty groups totalling over 20 per cent of the population. This best-seller led in a whole 'poverty literature', going on from the poverty of West Virginia and Harlem to discover the far more desperate conditions of the Indian reservations and the Mississippi flood plain. In 1968 the Citizens' Crusade against Poverty confirmed that some Americans actually starved, and CBS television filmed children as they died of hunger.

After the literature came the statistical research, and the introduction of poverty as a political issue. In 1960 Kennedy became President looking for a 'New Frontier', and in 1963 Johnson set out for 'the Great Society', winning a much greater electoral triumph in 1964 than even Roosevelt had known. Of course it did not last. The exhilaration waned, America sickened of foreign wars and domestic strife, and in 1968 opted to return to the quiet life. But poverty had undoubtedly been rediscovered; also something about 'opportunity' and 'participation'.

Johnson called the sixties the decade of opportunity, but they will also be remembered as a time of discontent and disappointment. The affluent society had lost its innocence.

The reasons why

Why did America rediscover poverty during a decade when it had never been wealthier, and when it attained its most rapid economic advance since the Second World War? Many authors have listed specific problems apparent by the late fifties. The growth rate, although sustained, was sluggish by continental European standards, and unemployment persistently stood above 5 per cent. Also, there were several lesser social disorders that could not escape notice. Juvenile delinquency was a growing concern in the big cities, prompting some of the first large-scale community action enterprises. The Kennedy administration augmented these efforts as part of the Delinquency and Youth Offences Control Act, 1961, and they were later merged into the far broader War on Poverty.

Furthermore, in this affluent society, the cost and caseload of public assistance continued to grow. A major cause was the apparent increase in illegitimacy, partly the result of the growing proportion of young unmarried adults in the population arising from the post-war high birth rate. This factor also influenced the delinquency problem. A minor 'welfare backlash' developed, distinguished by the 'Battle of Newburgh' (1961). Newburgh is a small town in up-state New York which decided to control its public assistance outlay by making the recipients queue for their money outside the town hall. However the news coverage was so hostile that the whole approach was discredited. A more positive approach was embodied in the 1962 amendments to the Social Security Act, reinforcing social work services to public assistance clients, although the ultimate objective was still to economize on the system.

These problems were in fact side effects of the increasing urbanization of America, sustained by industrial prosperity. In the long run the major social impact of this was, as we have noted, that internal migration threw into the limelight the deprivation of the less developed parts of the nation. It also precipitated a crisis in race relations.

Several authors attribute the entire concern about poverty to the black Civil Rights movement. This is consistent with the view that government responds to powerful organized interest groups, and Civil Rights in the late fifties was the only large interest group that even indirectly represented the poor. Yet the role of this movement, and black protest in general, in the rediscovery of poverty is very complex. The racial and poverty issues stemmed from the same situation, and had a powerful interaction, but most black movements were not principally concerned with economic affairs. The Black Revolution like England's Irish Question passed through three phases; legal emancipation, economic grievance, and nationalism.

The Civil Rights movement, as the name suggests, began as a campaign against the legal disabilities of black people in the South. Its supporters claimed the right to vote, to sit and eat with white people at the same lunch counters, to ride on the same parts of buses, to swim at the same beaches and to send black children to the same schools. The non-violent style of protest, particularly the sit-in, developed in this campaign in the 1950s, became the established fashion of protest for many other groups, particularly students, long after black protest had moved to other techniques. This phase took place under the traditional black leadership of the Baptist and Methodist clergy.

The original aims were largely realized by the Civil Rights Act of 1964, which banned all overt forms of racial discrimination except in housing. (The Southern town where I later lived promptly closed its three new civic swimming pools, which three years later were in an advanced state of dilapidation.) Housing discrimination, a sensitive issue outside the South, was outlawed in 1968. But just as the Irish passed on to agrarian reform, after ending legal discrimination against Catholics, Civil Rights now tried to turn itself into a Poor People's movement, culminating in the March of the Poor on Washington in

1968. However this was a far less impressive demonstration than the Civil Rights march five years earlier. The new issues were less clear-cut, and the economic problems of the Southern black peasantry had little relevance to the slums of the North, whose residents started to generate their own activities.

The city blacks, better fed and educated than their southern relatives and less responsive to traditional church leadership, passed rapidly to the nationalist phase, Black Power (c.p. Sinn Fein : 'Ourselves Alone'). The Black Power Movement rejected the alliance with the white liberals made by Martin Luther King in the 1950s, and even Ralph Abernathy's link-up with the white poor and the smaller minorities. It also preferred arson and looting to sit-ins, although as yet the level of violence has been very restrained by Sinn Fein standards. The Watts riot in Los Angeles in 1965 is usually taken as the start of this activity, with its high point in the Newark and Detroit disturbances of 1967.

These were partly economic riots, directed against the property of exploiting white landlords and small businessmen in the ghettos. They have had a practical success in advancing 'black capitalism'. The more recent planned assaults on police, so far isolated incidents, represent a more strictly political protest (the final sequence of the Irish troubles began with shooting policemen). Whether this will, or could develop into a full separatist nationalism is doubtful, since it has no clear territorial base. It might be more accurate to see Black Power as principally a cultural movement, appropriate to the second generation of a migrant population exploring its identity in the new society. But there is also an element of class consciousness, since white poor do sometimes join in black riots, and there may be further developments in this direction.

Thus, black protest helped indirectly in drawing attention to the poverty issue, but was not itself very concerned with economic issues, until well after the main federal anti-poverty schemes were launched. Its main influence was probably on the development, rather than the origin of anti-poverty efforts, biassing them towards the black poor, sometimes at the expense of poor whites.

Nathan Glazer, speaking at a national conference on poverty in 1965 (Gordon, 1965), attributed the attack on poverty to a small group of liberal intellectuals in positions of power. The Administration now had a will of its own, able to mould public opinion rather than being moulded by it. This explains the greater progressive zeal of the levels of government farthest removed from the people, and recognizes the great personal contributions of Kennedy and Johnson.

It was indeed a time of reform from above, with the Administration fostering enthusiasm rather than being swayed by it. But we must admit that the political climate was at least permissive of such action. Whereas there were no strong material interest groups ranked on the side of the poor, there were none opposing them. As first conceived, anti-poverty legislation attacked nobody's interests. The Administration built on a broad, disinterested, ideological support, nourished by a genuine sense of guilt that poverty was possible in a nation so rich and so vocal about freedom and opportunity.

The political history of the sixties spans the rise and fall of the reformist Democrats. Kennedy ruled with a rather unco-operative Congress, and achieved few tangible reforms. Yet the manner of his death left Johnson with a great fund of popular and congressional good-will which was capitalized in the 1964 election. This was also the time when the Administration embraced Civil Rights.

From 1964 to 1966 anything was possible, even a mild degree of 'socialized medicine' in the Medicare scheme, and federal aid to local school systems. After the 1966 Congressional elections the pace was slower. Vietnam drained off much liberal emotional energy, more important than the resources it consumed. The dispirited President decided not to run for office in 1968. The urban problem became an issue of law and order more than community organization, and in 1968 the administration sank. It only just sank, and but for the fiasco of the 'police riot' at the Chicago party convention it might have wallowed on. It might even have been buoyant again had Robert Kennedy lived to lead it into the seventies—though it is possible that he might have torn it apart. As it is, some of the enthusiasm raised during the decade is still alive on the college campuses, in the community action agencies, among social workers, and in its own way among the black militants and even the hippies. What has died is much of the faith that further effective change is possible in the present structure of politics and of society. The sixties raised more expectations than they satisfied. What does this mean for the next decade?

The British experience

An Englishman cannot escape the parallel with his own country. We shared the affluent society, at about half the American standard of living; luxury enough in comparison with the austerity of the forties. Poverty was considered dead. Rowntree's third survey of York (1950) seemed to prove this, at least by the standards worked out in the previous (1936) survey. Social workers concentrated on personal inadequacies. Even the standard university text, Penelope Hall's, *The Social Services of Modern*

England, discussed primary poverty solely as a pre-war problem.

Then the poor crept back into the academic and literary light. They came first as gypsies and homeless families, then as more normal groups, large families and the aged. In 1965 Abel-Smith and Townsend asserted that over 14 per cent of the population was in poverty, with a cut-off at about £700 p.a. (app. $1,700) for a family of 4 (*The Poor and the Poorest*). This was the usual national assistance standard at the time, including rent and other allowances and 'disregarded' income. The break-through in the popular media was the TV play *Cathy Come Home*, first shown in 1966 and twice repeated. This was Britain's mild equivalent to CBS' *Hunger in America*.

In 1964 Britain lost its Conservative government by a narrow margin. Eighteen months later it re-elected the Labour Government with overwhelming support. There was talk of social reform, family allowances were increased and the insurance system will be re-cast. But possibly the Labour Government has already met its Vietnam in the economic crisis. Even while the reforming zeal of some groups still increases, the popular climate is becoming less accepting.

The same pattern, with modifications, is visible in several West European nations. For instance De La Gorce's *La France Pauvre* (1965) is directly modelled on Michael Harrington, and draws similar conclusions. Each nation experienced some form of social revolution after the war, and spent a decade thinking that its society had been remade close to perfection. By the sixties the generation that shaped the post-war world was close to retirement—with some conspicuous exceptions. The new dominant generation was less committed to the system, and better able to see its shortcomings. Also, even in Europe the affluence of the fifties created problems of rapid urban growth (after all why had Cathy no home?), internal migration and race relations. Many of the problems of America's cities have their echo in urban Britain.

6
The strategies of war

The growth strategy

The anti-poverty legislation of the Democratic administration can be considered under four strategies, reflecting different ideas of the main causes of poverty. These four approaches were pursued simultaneously, but the emphasis tended to shift during the decade.

At the outset poverty seemed to be a problem of laggard economic growth to be solved by stimulating demand. The rapid increase in population masked the fact that the per capita growth rate in America in the fifties was as low as in Britain, and much lower than in most of Europe. Unemployment had been about 5 per cent for most of the time since the Korean War. The new Administration responded with a large budget deficit, and for the first time in America managed to get such a piece of Keynesian economics whole-heartedly approved by the business community. Keynes no doubt would have preferred the deficit in the form of an increase in public spending, particularly directed to the lower income groups. This would pass on the whole weight of purchasing power, absorbing little in savings, and might seem especially appropriate as an anti-poverty measure. In practice, America increased the purchasing power of the rich with a massive tax cut, proposed by Kennedy in January 1963 and passed by Johnson in February 1964.

Yet already the economy was stirring, and, helped on by the tax cut and the Vietnam War, the nation swung into the longest spell of sustained business expansion in its history, its highest growth rate and lowest unemployment rate since the World War, and its sharpest spasm of inflation. By 1968 Johnson had to impose a 10 per cent tax surcharge to avert an exchange crisis. The expansion began before the major American commitment to Vietnam, but it was supported and intensified by the war. Ironically the war which discredited the Administration was probably its most effective anti-poverty measure.

Even so, general expansion could do little for the more backward areas. Accordingly, the Area Redevelopment Act of 1961, and its successor the Economic Development Act of 1965, concentrated federal funds on high unemployment districts. The emphasis was on grants and loans for public works. The Appalachian Regional Development Act of 1965 directed further funds at an eleven-state region, focusing on growth districts without any unemployment means test. This was a federal-state programme, with the main effort in road building. However 25 per cent of the funds was reserved for 'social' projects, such as health facilities and schools. Like the tax cut, these schemes gave no aid directly to the poor, but relied on grants to business and local government to 'trickle down' to the unemployed.

The employment strategy

Increasingly this percolation seemed inadequate. Shortages of labour developed even while some groups still showed little increase in employment, notably black people and teenagers. Boeing of Seattle ranged America and Europe for aircraft workers, while in the city's black ghetto 7 per cent were unemployed. Hence the 'employment strategy' aimed directly at the unemployed, particularly the hard core.

The best known employment measure is probably the Manpower Development and Training Act (MDTA) of 1962. This was inspired by unemployment resulting from automation, and envisaged retraining experienced workers in new skills. Federal grants went to employers, state agencies, and trade unions for in-service training, and to state and private establishments (principally the Community Colleges, equivalent to Britain's Technical Colleges) for vocational education. These schemes included subsistence allowances for trainees, and courses in basic literacy. The Act was renewed several times, but the original intention to phase down federal support in favour of local financing has been abandoned.

But to train a man who already has considerable industrial experience is one thing: to reach the unskilled immigrant from the South, the unmarried mother, the drug addict, or the man with a police record, is very different. In 1966 the MDTA programme was reorientated to put more emphasis on training the disadvantaged, that is, the long-term unemployed, school drop-outs, black Americans, and older workers, with some success. However there are still groups that MDTA cannot reach.

The Concentrated Employment Program (CEP) of 1967 is an attempt to extend the reach of MDTA. This attempts to create new jobs for the disadvantaged, with a federal guarantee to employers to improve the employability of certain workers, through training, day nursery services, health services, or whatever is required. Each guaranteed worker is served by a counsellor to guide his progress until he is established. The scheme is operated by MDTA and community action agencies.

CEP was soon joined by JOBS, Job Opportunities in the Business Sector. This is a similar scheme, in which private enterprise finds the jobs and the federal government meets the additional training costs. JOBS is promoted by the National Alliance of Businessmen, under the chairmanship of Henry Ford II. It canvasses for job openings in 50 target cities, and has pledged 100,000 jobs by June 1969. A companion programme, PACE (Public and Community Employment) has recently evolved from the New Careers scheme developed by community action agencies. So far this concern for training and rehabilitation has created a bewildering variety of schemes, novel and ambitious by British standards, but with nothing as comprehensive as Britain's 1963 Industrial Training Act.

The Nixon administration has taken up the renewed enthusiasm for the hard-core unemployment schemes shown in the last year of the Johnson regime, but, since it is also committed to controlling inflation by cooling off the economy, the future of the whole employment strategy is obscure. In particular, the latest training efforts have all pre-supposed a labour shortage.

The Civil Rights Act was also an important part of the employment strategy, aimed at ending job discrimination. It has perhaps opened up some jobs to qualified black workers, but still the black unemployment rate has never ceased to run at twice the white rate (it is of course arguable that a fairer treatment of black applicants would simply shift the burden of unemployment to the hard-core, white poor). These anti-discrimination laws have been followed by positive efforts to promote 'black capitalism', which is presumably less discriminating than the white variety. Most of these enterprises are co-operatives, and have achieved some success, particularly in Philadelphia.

Educational schemes can be viewed as forming a separate anti-poverty strategy, or as a sub-category of the employment strategy, since most of them emphasize increasing employability. Until 1965 federal intervention in education had been crippled by constitutional problems about government aid to church schools, since a subsidy to poorer districts could scarcely ignore the Catholics. The federal government had been forced like Santa Claus to do good by stealth, using defence Acts to subsidize schools near military bases, defence-related school subjects, and ex-service students. But finally, in the Democratic heyday of 1965, Congress passed the Elementary and Secondary Education Act (ESEA). This offered grants to all schools in poor districts for remedial programmes for educationally deprived pupils. The main priority has been on remedial reading, but local educators have a wide discretion in the type of proposals they can offer for funding. Probably the greatest benefit of this programme has been to focus the attention of teachers and administrators on the problems of educationally backward children. ESEA projects have probably reached more of the poor than any other anti-poverty scheme, but unfortunately the long-term effect of the typical short crash courses seems to be slight. After an immediate improvement in the pupil's performance, the influence of their environment gradually re-asserts itself. They tend to leave school just as soon and no more literate than before.

Poverty as a pathology

The employment strategy shades off into the 'curative strategy'. This is the main province of the professional social scientists and social workers. It recognizes that the hard-core unemployed have a complex of problems which job-finding, vocational training and formal education cannot by themselves overcome. Most of the federal curative programmes are contained in the 'War on Poverty', and will be described in the remaining chapters of this book. The main efforts outside this structure fall to the public assistance departments, and were laid down in the 1962 and 1967 amendments to the Social Security Act.

Both these sets of amendments concentrated on amplifying the services delivered to public assistance clients, turning what was almost entirely a cash payment system into a serious social service structure. By the late sixties this had led to frequent specialization within public assistance departments between 'basic services' (cash) and 'social services' (social work), with the professionally trained social workers concentrated in the latter. This is quite contrary to the tradition of American social work that cash payment is a service, and it brings America closer to the British concept of cash assistance as a fairly impersonal administrative function.

Most of these services are concentrated on AFDC mothers to make them self-supporting. The main provisions are training schemes, day nurseries and family planning advice. Social work intervention, through case and group work, is to motivate the mothers to seize these opportunities. Congress specifically recommended the states to cut off assistance to clients who did not respond.

The curative strategy is strongly coloured by the concept of the Culture of Poverty. This concept stems from the anthropologist Oscar Lewis, working among the uprooted, unemployed peasantry swarming into Mexico City (Lewis, 1959 and 1962). He saw it as the pattern of behaviour by which particular groups of the poor cope with the hopelessness and despair of their environment. The salient features are a lack of participation in the major institutions of the wider society, a minimum of social organization beyond the family, unstable family patterns, and a strong feeling of 'marginality', reflected in fatalism and a low level of aspiration; 'an inter-generational transmission of values and practices that inhibits constructive action'.

Lewis doubted whether such extreme demoralization existed among the United States poor, but his vivid case studies, consisting often of edited transcripts from tape-recorded interviews, made the concept well known and widely applied throughout the USA. Harrington used the concept liberally in relation to North American groups, and laid great stress on the psychological distinctiveness of the poor.

The concept is attractive in that it seems to exonerate the wider society, and treats poverty as a symptom of deviant behaviour. America should continue with its traditional task of acculturating an 'alien' minority. This is reflected in the two shilling booklet, *Low Income Life Styles,* distributed by the federal Welfare Administration in 1967. This is a digest of all the major US studies on this topic, and concludes of the poor that:

> 'while they accept typical American values they are frequently lethargic in trying to attain them.
>
> 'Energetic patience must prevail. The alienated adult cannot be completely re-educated. His children can be somewhat swayed. But it is with his grandchildren that one can really have hope.'

This might be described as the theory of the self-defeating poor. However, other commentators have inverted the concept, stressing the role of the institutions which impinge on the lives of the poor. Public assistance, schools, police, courts, public and private landlords and even voluntary social work, the retail trade and the telephone service, all socialize the poor to accept deprivation. The poor are more victimized than self-defeating, and the remedy lies in institutional change rather than client motivation; in redistributing power rather than opportunity. It is often obscure which concept, victimization or self-defeat, activates many curative programmes. It depends which audience they address—the institutions prefer one theory and the poor the other.

Incidentally, the Culture of Poverty and its variants have been noted at other times in history:

> 'The failings of the workers in general may be traced to an unbridled thirst for pleasure, to want of providence, and of flexibility in fitting in to the social order, to the general inability to sacrifice the pleasure of the moment to a remote advantage. But is this to be wondered at? . . . The social order makes family life almost impossible . . . In a comfortless filthy home no domestic comfort is possible . . . society daily and hourly commits social murder.' Engels, *Condition of the Working Class*, London 1844.

Direct aid—alleviation or cure?

> 'It's great stuff this War on Poverty! Where do I surrender?'
>
> (*Public assistance client, 1966.*)

One way to eliminate poverty might be to give the poor money. The 1935 Social Security Act had already established a twofold system, assistance and insurance, to give away federal money, and there was the possibility of opening up other channels, such as the Internal Revenue. However, the extension of direct aid has tended to be the least popular anti-poverty strategy in America. The continued existence of poverty was often taken as proof that the 'New Deal' approach was no longer appropriate, rather than that it should be pressed more forcibly.

Of course this is an expensive strategy, at least for the higher income tax-payers. Naturally, for the nation as a whole internal transfers of income cost nothing. Yet even as a tax-raising problem, the financial elimination of poverty is not very formidable. Orshansky estimated that the amount needed to bring every sub-standard family up to the poverty line—bridging 'the poverty gap'—in 1963 was about $12,000 million. This was about 2 per cent of the national product, or about half the cost of the Vietnam war in 1968. However, wars, against poverty or the Vietcong, are one day expected to be won and finished, not to be an endless charge. The object of federal policy was to eliminate poverty by turning 'tax-eaters into tax payers'. Also, any other objective would do permanent violence to the 'normal' wage-price structure of society, and America's achievement-orientated ideal.

Apart from the cost, there is a more theoretical argument as to whether money grants can actually cure poverty. It has been claimed that this merely 'rubs money into the sore', reinforcing the apathy of the recipient, and securing him in his culture of poverty. Ultimately, this argument has to define poverty in something other than money terms, seeing it more as a lack of the chance to 'achieve' in society. Several slogans have been coined to support this view, 'Poverty is a social not an economic problem', and 'We have proved that giving money is not enough'. However this assumes that the poor have actually been given enough money, which is debateable. The counter-argument is that a substantial change in income, more than public assistance usually attempts, really can change the life-style of the recipient. Even if it does not, a large part of the poor must and should be accepted as dependants. 90 per cent of public assistance recipients are classified unemployable, mostly because they are too old or too young to work, or have urgent family responsibilities. Must we try to make everybody employable?

The biggest measure of direct aid under the Kennedy-Johnson administration was probably Medicare (health insurance for the elderly), contained in the 1965 amendments to the Social Security Act. Its companion measure in public assistance was Medicaid, a potentially wide-ranging scheme, but which can be interpreted by different States in many different ways. There have also been several increases in social insurance benefits, the largest of which in American history came into effect in January 1968. Finally, public assistance has been liberalized (1962) to include families with unemployed fathers in the AFDC category at the discretion of each state (most states have not yet used this provision) and the distribution of free food—in kind or in tokens—has been added to the PA function. Very limited use has been made of housing aid, although there was some revival of interest in 1966-8. FHA mortgage insurance is now less restrictive, rent subsidies are available for tenants in non-profit housing schemes, and there are some low-interest loans for house purchase and improvement. Finally the Model Cities Program has evolved out of earlier slum clearance projects, and is being spread (more thinly?) by the Nixon administration across most of urban America.

The boldest direct aid scheme is still in the early experimental stage. If nationally adopted, it would give America a more sophisticated income maintenance system than any yet existing. It is the Negative Income Tax (NIT), publicized by both extreme Right and Left wing theoreticians (e.g. Milton Friedman and Robert Theobald) and accepted in 1968 for a pilot project with 1,000 families in New Jersey. The popularity of this idea came on very rapidly in the late sixties, and represents a partial conversion of America to the Direct Aid strategy.

The scheme uses the income tax system to assess everybody's individual means and needs, as at present, and then it either taxes or makes grants as required. It would function principally as a very flexible selective family allowance system. Indeed it is seen largely in lieu of family allowances, which have limitations as an anti-poverty device. America is not interested in family allowances as a population incentive, a consideration which dominates the structure of many European family policies.

The pilot scheme is conducted by the Office of Economic Opportunity, once dedicated to the thesis, 'A Hand-Up, not a Hand-Out'. It would be curious if OEO's main contribution to human welfare turned out to be a resurrection of the Speenhamland System, the early form of NIT which England abandoned in 1834 in favour of the workhouse.

7
War declared

'... We have never lost sight of our goal—an America in which every citizen shares all the opportunities of his society, in which every man has the chance to advance his welfare to the limits of his capacities ...

The distance that remains is the measure of the great unfinished work of our society.

To finish this work I have called for a national war on poverty.

Our objective—total victory ...

Because it is right, because it is wise, and because for the first time in our history it is possible to conquer poverty I submit the Economic Opportunity Act of 1964.'

(President Johnson's message to Congress, 16 March 1964.)

'The overtones of much of the discussion and its sudden eruption create the impression that we are in the throes of another emotional jag.'

(Professor Margaret Reid, 24 April 1964.)

With this grandiose rhetoric the President christened the provisions of the Economic Opportunity Act of 1964 'the War on Poverty'. However, most of the Administration's domestic policies throughout the 1960s attacked poverty. The EOA programmes are actually one of the least important and least expensive campaigns in a much wider war, taking about a tenth of the total federal anti-poverty expenditure (see Table I).

Yet this was the scheme, hurried together from scratch in the emotional few months after President Kennedy's murder, that was to focus the sentiment of the nation and bring on 'another emotional jag'. It deserves our attention as the most remarkable and uniquely American part of the anti-poverty effort.

The Act was a curious package of ten largely unrelated schemes; a microcosm of the entire anti-poverty efforts of the decade. Each scheme would have been more logically included in other Acts, such as ESEA, the regional development schemes, the social security amendments, or the various employment and training acts. The EOA was partly an excuse for the President to push through a collection of programmes, several of which had been before Congress previously, thrown together under an exhilarating title. Johnson had snatched the standard of the fallen Kennedy, and now rallied the nation with a new call to arms.

The one unifying feature of the EOA programmes is that they were all funded through the Office of Economic Opportunity (OEO). This was set up in the Executive Office of the President, outside the regular departmental structure of the federal government. However, most EOA programmes were delegated out to established departments, with OEO in a limited co-ordinating role. These are the more conventional schemes, which we will summarize first.

TABLE I

The Federal Anti-Poverty Effort, Fiscal Year 1964-5

(The American Fiscal Year begins on 1 July. Thus 'Fiscal Year 1965' refers to the year ending 30 June 1965.)

The federal anti-poverty programmes represented 3½ per cent of the current gross national product.

A. Estimated federal expenditure outside OEO=$18·378 billion (a US billion=1,000,000,000)

1. *Cash Payments*

   OASDI
   Unemployment insurance
   Public assistance

   Total=$14 billion

2. *Education*

   Aid to slum schools
   Teacher corps
   Aid to impacted schools
   Federal scholarships
   Manpower training

   Total=$2 billion

3. *Food*

   Food stamps
   Free commodities

   Total=$268 million

4. *Housing*

   Rent subsidies
   Public housing

   Total=$100 million

5. *Regional Development*

   Economic Development
   Appalachian programmes

   Total=$2 billion

B. Estimated state, local and private philanthropic expenditures =$15 billion

C. OEO expenditure (1964-5 was 1st year of operation)=$800 million

(Estimated OEO expenditure for 1968-9=$1·9 billion)

Note: The definition of anti-poverty expenditure is extremely arbitrary. Many welfare programmes, such as ex-servicemen's benefits, and public health services, are not included above, although normal social insurance payments are. The programmes listed above are those primarily envisaged as anti-poverty schemes, although many other programmes also relieved hardship, and not all the recipients of social insurance benefits were poor.

Delegated programmes

These are mostly employment programmes or investment incentives for the self-employed, forming part of the 'employment-education' strategy. The programmes directly operated by OEO tend more towards the 'curative' approach.

The *Work Experience* programme was in the tradition of enlarging the services available to public assistance recipients. It was delegated to the Welfare Administration of the Department of Health, Education and Welfare (HEW), which supervised its actual operation by the state public assistance systems. Reminiscent of some Elizabethan Poor Laws, it aimed at giving activity to the long-term unemployed, to restore them to the habit of work. Initially these tended to be make-work programmes of the leaf-raking type, but later, in some states, a stronger element of training and useful experience emerged. However this was still mainly in public open spaces, graduating from raking the leaves to tending the plants. The 1967 Social Security amendments introduced a *Work Incentive* programme, administered by the Department of Labor operating through the state employment services. This is a more conventional training scheme, with special attention to AFDC mothers, linked with public assistance day nursery services. It is gradually being expanded to replace the Work Experience programme which is now almost phased out.

The other training schemes concentrate on teenagers, because this is the major area of unemployment, because in terms of the 'culture of poverty' it is a good point to break the 'poverty cycle' (the transmission of the culture of poverty between generations), and because the schemes can be an adjunct to formal education. *The Neighborhood*

Youth Corps (NYC), delegated to the Department of Labor, creates part-time and holiday jobs for secondary school pupils from poverty-level families, so that they can stay on at school. My own view is that it is a poor substitute for educational maintenance grants, diverting too much of the pupils' time (up to 15 hours a week during school term) into routine, unrewarding activities, such as cleaning offices or washing laboratory equipment. However, since there are no maintenance grants, this is the best alternative. It is also supposed to give the recipients more self-respect, and the work they do should ideally motivate them to higher education and socially useful careers.

The federal grant goes to non-profit agencies of any type, usually public services, to employ eligible teenagers on work which does not displace regular employees. This forces a great strain on the imagination of the agencies, to think up new tasks that teenagers can competently discharge. I have myself worked with a similar scheme in public assistance which I believe was imaginative, although most of the ideas originated from the teenagers—home tutoring, dressmaking classes, holiday excursions for low-income children, children's playgroups, and so on; but the NYC seems often to be used to carry out routine chores, which otherwise would be performed less often for lack of funds. There is also an NYC scheme for drop-outs, providing 23 hours a week employment, plus 8 hours of training, education, and counselling, with the hope of returning many corpsmen to school. Since 1966 certain commercial enterprises have also been able to take NYC recruits for skilled training, recover training costs, but not normal wages, from the federal government. There were about half a million NYC openings in July 1968.

The *Work-Study Program* is an NYC scheme for college students. Universities and colleges receive the funds from

HEW, recruit eligible students, and cultivate suitable employment in local non-profit agencies. Most of the jobs are no more educational than any other student employment. Like NYC, Work-Study exists in lieu of student maintenance grants, and employment is limited to 15 hours a week. It is useful partly because other federal programmes to employ the disadvantaged have narrowed the demand for student labour. The original intention, to restrict the scheme to students from poverty-level families, was frustrated by the scarcity of such students on American campuses. Hence, it now operates on a much broader definition of low income, and has severed its connections with OEO (1966).

Upward Bound, is a related campus programme. Until July 1968 it was run by the Community Action section of OEO, but it has now been transferred fully to HEW. The scheme aims to inspire talented low-income school pupils who are in danger of 'dropping out', to complete school and continue to higher education. These 'talented risks' are identified by the high schools, and referred to the local university for a two-month campus vacation course, in the last summer before school graduation. This stimulus is followed by a tutoring programme throughout the final year of school, using Work-Study student tutors, and a second pre-university two-month course. About 300 such schemes exist, with about 25,000 students, 80 per cent of whom are expected to enter university. Financial assistance is available. The main drawback of the scheme is that it often succeeds too well, making high school work seem even more tedious and the university first year rather an anti-climax.

The 1964 Act spanned the whole range of America's educational problems, from Upward Bound, which presses secondary education to its extreme limits, to an

Adult Education programme to eliminate 'functional illiteracy'. This latter scheme was delegated to HEW, which took over full control in 1966. It allots funds to state education agencies in relation to the proportion of the adult population in each state with less than six years' schooling.

The investment incentives include the *Rural Loans* programme, delegated to the Farmers' Home Administration of the Department of Agriculture. It has since (1966) become the full responsibility of this department, although OEO continues to help in screening applicants. The original intent was to help small family farms to buy more land, but its most conspicuous success has been to promote rural co-operatives and various non-farm activities. The urban equivalent is the *Small Business Loans* scheme, to help set up small retail and service enterprises. It has also been used to promote co-operatives. These schemes are aimed to give small business its share in federal economic development policies. More comprehensive efforts for neighbourhood economic development have since evolved in the Special Impact programmes operated through the community action structure.

Direct programmes

OEO itself is directly responsible for four programmes, Migrant Farm Workers, Job Corps, VISTA and the heart of War on Poverty, the Community Action Program (CAP).

The *Migrant Workers* scheme channels OEO funds to voluntary and statutory agencies to provide housing, sanitation, education and day nurseries for some of America's 2½ million migrant farm workers and their families. One of its main efforts has been in educational work, to ease the shift of this labour force into other employment in the face of increasing farm mechanization.

Job Corps is basically a revival of Roosevelt's Civilian Conservative Corps of the 1930s, and had been unsuccessfully put to Congress several times before. The idea is a series of residential camps, mostly in the wilder, remoter parts of America, where 'alienated youth' can go through character-forming experiences to become good and useful citizens. The Corps has probably attracted more criticism than any other anti-poverty programme because it is very expensive in relation to the numbers it serves, most of it is not devoted to direct training in marketable skills, and its clientele does not attract public sympathy. Many Congressmen resent 'coddling thugs', and scandals in the Job Corps can always get publicity. Undoubtedly the Corps has generated some interesting news stories. One absconding Corpsman stole an aircraft and crashed it in mid-Atlantic; there have been some well-reported fights and occasional knifings; and the girls' camps have been well quarried for tales of teenage promiscuity. Not surprisingly, communities often object to JC camps in their districts, and relations with local populations are sometimes difficult. Such hostility is a definite handicap in rehabilitating young people, who by definition are alienated, and a great deal of effort is now being put into community relations.

Job Corps applicants must be teenagers, out of school, unable to find or hold adequate jobs, and from an 'underprivileged' environment. It is a voluntary scheme and most applicants are referred by the state youth employment services. Many entrants have delinquent backgrounds, but on the whole JC get 'the cream of the disadvantaged'. Without compulsion, those hardest to reach will always be last reached.

The 82 Conservation Centers are the basic element of the Corps. Located in wilderness areas, with 100 to 250 enrollees in each camp, they offer strenuous outdoor activity, training in forestry and related skills, basic education, and instruction in nutrition, hygiene and physical fitness. This usually lasts nine months. The ethos is rather like an English boarding school, even to dressing the corpsmen in dark blazers with gaudy badges. There are elements of a prefectorial system, an elaborate graduation of small privileges and minor rewards, lots of team games, and plenty of shower baths.

From Conservation Centers or as direct enrollees, corpsmen may enter the 6 Urban Centers which train for specific vocational skills. These are big camps, often on disused military sites, with 1,000 to 3,000 enrollees, operated under contract by business companies, educational or social agencies, and universities. These centres have expanded greatly since 1964, since their utility is more directly obvious to Congress than the Conservation Centers.

The original draft of EOA did not provide for Women's Centers, but they were included in a Congressional amendment. The proportion of female enrollees has since risen steadily. The 18 Women's Centers are similar to the Men's Urban Centers in that they are accessible to towns, and emphasize vocational training, but they are smaller and incorporate some of the boarding school aspects of the Conservation Centers. In addition to basic education, the women's programme features home economics and grooming, to socialize the girls into an acceptable female role. These centres are also managed under contract.

In March 1969 the federal General Accounting Office released a two-year evaluation of OEO-financed programmes. While approving of most programmes, it sharply criticized the Conservation Centers for the disappointing post-Corps employment experience of their 'graduates'.

However it recognized that 'Corps members have also received benefits . . . such as health and social and psychological development, which are generally not subject to measurement'.

In July 1968 there were approximately 23,000 male and 10,000 female Job Corps enrollees.

Volunteers in Service to America (VISTA) is a domestic version of Kennedy's Peace Corps, with a more radical bent. It uses volunteers mostly on a one-year contract to serve in any aspect of the anti-poverty effort. It has an initial six weeks' residential training course and its own journal to help promote a strong *esprit de corps*. Recruiting teams of ex-volunteers ('return VISTAs') hunt for new recruits on most university campuses, and recruiting literature goes out to old age pensioners with their benefit cheques. Although it tries hard to attract older volunteers, and it has had some outstandingly successful 'senior citizens', it remains mostly a young people's organization. Most volunteers are college students, although many have not yet graduated but are interrupting their studies. Foreign volunteers are fairly common, since unlike the Peace Corps there is no citizen requirement. Recently there has been an experiment at the University of Maryland to combine VISTA experience with post-graduate training in social work.

VISTA is the modest Red Guard of the War on Poverty, searching after a modest little grass-roots anti-bureaucratic Cultural Revolution. It attracts non-conformist middle-class youth, and the training course and the moral fervour and indignation of *VISTA Volunteer* foster the sense of Crusade. To most volunteers the poor are victims of society, and they envisage themselves reforming conventional welfare agencies, or spearheading militant community action. The first volunteer to be assigned to a public assistance department (it was rather adventurous in asking for a VISTA) confessed that she was dismayed at being sent to the 'arch-enemy'.

This attitude creates problems. Several agencies have found VISTA an embarrassment. Yet this spirit sustains their morale in lonely, trying situations, with few physical comforts. In November 1967 there was even a complete halt of funds, when OEO ran out of money before the new financial appropriation had been approved, and Congress, debating whether to renew OEO, refused to bring in a supplementary grant. Salaries and maintenance allowances in every OEO scheme stopped for a month, and some workers had to board with the families they served. 'I think it's great, because now some of the volunteers will have to get out in the community or starve' wrote one VISTA ('The Day the Money Stopped', *VISTA Volunteer*, December 1967).

The training course is a series of mildly traumatic experiences. At the centre I attended for a few days ('some people here say you're from the CIA' confided one trainee) the course included a week's residence with a poverty level family, and the 'dropping off' exercise. Trainees are 'dropped off' in a strange town with $20 and the instruction to discover as much as possible about the social structure of the community in four days. Speeches from extremist politicians of contrasting views, including Black Power invective (most of the volunteers are white) are part of the diet. Even spells of boredom when the programme arrangements collapse are considered educational —'You'll have to sit through lots of boring committee meetings if you join a Community Action Program!'

By 1968 about half of all VISTA assignments had been in about 200 rural projects, mostly in the South and Appalachia. Florida, West Virginia and Kentucky lead in the use of volunteers. Teaching and tutoring were the main activities, but they also helped with co-operatives, clinics, information bureaux and recreational programmes.

About a quarter of the VISTAs have worked in urban slums, mostly with Community Action Agencies. Teaching is again the main activity, followed by organizing group action of many varieties. VISTAs are barred by law from political activities, but they have been linked with militant movements such as rent strikes (New Jersey, 1965).

About half the Indian reservations in America have had VISTAs, notably the Navajo reservation in Arizona, which had over 50 in 1968. Volunteers have been widely employed in mental health projects, and to a limited extent in Job Corps centres (about 60 by 1968). In July 1968 there were just over 5,000 serving volunteers.

The argument for volunteer effort is that volunteers can bring a fresh approach to problems and break through traditional attitudes. They may be more acceptable to 'hard-to-reach' clients, and the experience is good for the volunteers and helps to sharpen the social conscience of America. Also they will do jobs that do not attract professional staff. Yet the fact that the volunteers mostly come from middle-class backgrounds might be considered to diminish their understanding of poverty problems and their personal acceptability to clients. Many community action programmes now emphasize the use of local low-income volunteers, and have discontinued VISTA. However, an illiterate volunteer can scarcely teach literacy classes.

TABLE II

OEO Budget Activity, Fiscal Years 1967, 1968, 1969 and 1970.

The figures for 1970 represent President Nixon's Budget proposal. This may be reduced by Congress, entailing some reapportionment between programmes.

PROGRAMMES	*1967*	*1968*	*1969*	*1970*
Work and Training	(in millions of dollars)			
Neighborhood Youth Corps	376	291	342	310
Job Corps	209	282	280	170
New Careers	17	8	19	—
PACE	—	—	—	25
Concentrated Employment	49	74	83	83
JOBS	—	60	152	307
Department of Labor Support	11	18	20	18
Special Impact (neighbourhood economic development)	7	20	22	46
Work Experience	100	44	10	—
Foster Grandparents	10	10	9	9
TOTAL	779	807	937	968
Other Community Action				
Head Start	349	318	318	338
Head Start Follow Through	—	15	30	60
Comprehensive Health Centers	51	33	60	80
Family Planning	4	9	13	15
Emergency Food and Medical	10	13	17	20
Legal Services	25	36	42	58
Senior Opportunity Services	—	3	4	6
Local Initiative (i.e. 'versatile')	278	321	332	332
Support (research, pilot programmes, technical assistance, etc.)	61	84	85	94
Upward Bound	28	32	30	—
TOTAL	806	864	931	1,003
Migrants	33	25	27	25
Rural Loans	24	17	6	—
VISTA	26	29	32	37
General Administration	14	14	15	15
Miscellaneous Transfers, etc.	6	17	—	—
GRAND TOTAL	1,688	1,773	1,948	2,048

Note: The Concentrated Employment, JOBS, PACE and New Careers are supported jointly by MDTA, Department of Labor, and OEO funds. Certain NYC and New Careers projects are included in JOBS and CEP prior to 1970. PACE combines New Careers with a programme modelled on JOBS.

8
Maximum feasible participation

Community action—crusade or service?

Community action was a technique fashioned, partly by expatriate Americans, for use in the underdeveloped nations, where cultural change seemed an obvious requirement for material progress. The cultural explanation for poverty in America lead Americans to turn the technique to their own communities.

Briefly, community action, alternatively known as community development, hinges on inspiring a community to seek local self-help solutions to its problems, rather than waiting on outside bureaucratic intervention. The latter is considered less desirable because it is often clumsy and ill-informed, and the community is less receptive. In America the technique was married with the established native practice of Community Organization, which meant principally the co-ordination of city health and welfare agencies, mainly voluntary organizations (community funds, health and welfare councils, and so forth). This has been an uneasy marriage, for while community action aims at social change, accepting conflict as part of the process, community organization emphasizes co-operation, and works for the consolidation of the present system. The marriage is symbolized in the OEO community action insignia of an arrow and a bracket.

Throughout the brief history of the OEO community action programme these two strands of thought have been reflected in two separate types of activity, 'organizing the poor' and delivering services. In the eyes of many practitioners services were subordinate to the organizational task, for instance in this proposal for community centres in Minneapolis:

> 'The primary goal of the Citizens' Community Centers, Inc., is to bring about basic changes in the local system that causes people to become poor and to remain poor . . . Direct services . . . shall be used as temporary rewards for those citizens who not only need them but who will desire to organize, to cause changes in the institutional arrangements of the community.'
>
> (*CCC Inc.*, *1967*.)

However, the provision of services has usually been the major function, although, at least in theory, these have always had some further aim than meeting the immediate needs of the clients. R. H. Kramer in his four San Francisco Bay area studies (Kramer, 1969) remarked that here community action 'was perceived as a social movement (a cause) or a social service agency (a function), or both, thus constituting a curious blend of professionalism, bureaucracy, social action and reform'.

Community action was also a pioneer form of 'creative federalism'. This had been foreshadowed in the Mental Health Act of 1963, which funded local ad hoc organizations to set up their own community mental health clinics, rather than developing a government system. The 1965 Education Act and the 1966 Model Cities Program developed the same approach, but without including the state governments as intermediaries. In all these the principle was that the federal government supported local proposals for local action, avoiding the standardized grant-in-aid. This keeps control in federal hands, for the support grants are not unconditional, but it allows local diversity and initiative. It also achieves a more direct relation between the federal government and the target populations, by-passing the state and sometimes even local governments. It was a new way to manage an intractable sub-continent.

The 1964 theory of the community action programme was that OEO stood outside the encrusted bureaucracies of the regular federal departments, appealing directly to the people, especially the poor, for their ideas and insights into 'breaking the habit of poverty'. The office would then help finance any practicable enterprises. It was envisaged that successful enterprises would ultimately find permanent local funding, or achieve their goals and disband. The system was to be small, flexible, and temporary; a catalyst for social change.

Finding the community

But who are the people? Who are the poor? OEO was not prepared to accept the established state and local governments as sufficiently representative of the target populations—indeed to some observers they were part of the pattern of society which kept the poor alienated and impotent. Instead, each eligible community was invited to establish an independent Community Action Agency (CAA). This would seek recognition from OEO to become the sole channel for Community Action Program (CAP) funds in its area. A CAA could evolve its own programmes, or accept projects submitted to it, but only it could forward the proposals to OEO. OEO could refuse recognition to a CAA, or withhold funds from any project, but otherwise in principle it had no managerial control.

It was not expected that CAAs would normally operate programmes directly. They were planning, co-ordinating and evaluating bodies, supervising and funding the work of delegate agencies. Any non-profit agency approved by the CAA and OEO could be a delegate, whether it was an established concern or formed primarily to receive CAP money. A CAA could even set up delegate agencies itself, if no suitable organizations existed.

To get CAP funds, a community needed a recognized poverty problem and a recognized CAA. The extent of poverty was determined by OEO according to six standard criteria, including family incomes, unemployment rates and the rate of rejection for military service. At first there were no rules to determine if an intending CAA was 'representative'—merely that it should have 'the maximum feasible participation of the poor'. Gradually this came to mean that a third of the board members had to be poor, either by income or association (i.e. by living in a slum district or being closely identified with the poor in some way, such as the pastor of a slum congregation). In the 1967 amendments to the Act a three-part formula was set out; one third of the board to represent the poor, one third to represent the political and administrative structure, including local government and the statutory welfare services, and one third representing community interests such as business, trade unions, civil rights groups and voluntary welfare.

It was not enough for a board to find some tame poor people, add their names to the list of members, and hope they never attended meetings. The federals were liable to demand a poor people's election, which was embarrassingly difficult to organize. Philadelphia probably has the most elaborate CAA elections, complete with campaign literature, posters, parades, beauty competitions, polling booths and voting machines. In the first election (1965) 3 per cent of the target area electorate voted; it was 5 per cent in 1966. More often candidates are chosen and elected at public meetings, or else election forms are distributed door-to-door, to be mailed back to the organizers. No system has yet attracted more than a tiny fraction of the possible turn-out.

Vanishing autonomy

From the outset CAA autonomy has been eroded both from the local and federal ends. The state governors gained the right to veto any particular projects in their states, a power used on several occasions, as well as the funds to give technical assistance. However California's attempt to take over direction of the state's CAP activities was sternly rebuked by the OEO western regional director. This would 'dilute and dissipate the local determination on which the war against poverty was built' (Lawrence Horan, May 1968).

Local government control of CAA affairs was also strenuously opposed, even to the extent of refusing recognition to some nationally famous city enterprises. Philadelphia, for instance, had operated an inter-agency community action scheme on a Ford Foundation grant since 1962, one of the prototypes for the OEO plan. In 1964 OEO refused recognition until the mayor reduced his appointees to one third of the board. Some mayors publicly asserted that OEO was 'trying to wreck local government by setting the poor against City Hall' (Resolution proposed at US Conference of Mayors, July 1965).

In 1967 the mayors captured the system. The Congressional amendments of that year required all CAAs to be approved by the county or city authorities concerned, which could alternatively nominate themselves as CAAs. Most of the thousand or so CAAs existing in early 1968 were approved, but community action was now part of the political establishment. The federal Administration was forced into this to buy Dixiecrat support in a less friendly Congress, and fend off the complete abolition of OEO.

Congress also disliked 'walking-about money'. Initially all CAP funds were 'versatile', for use on any approved projects. Gradually Congress began specifying the use of grants, until by 1967 two-thirds of the funds were earmarked. The earmarking was often for standardized projects, which OEO itself began to develop; the so-called 'canned programs'. The first and greatest canned programme was 'Head Start', launched in January 1965 with Lady Bird Johnson as its honorary chairman. This is a nursery programme for infants from low-income families, and by 1968 it consumed a third of all CAP money. With the funds comes a bundle of federal guidelines, laying down the hours of class activity, teacher-pupil ratios, nutrition and health standards and much else. No earmarking was included in the 1967 Congressional amendments, but this was on the understanding that it would be continued administratively. A major consequence of earmarking is that the federal government can now swing the emphasis of community action to whichever anti-poverty strategy it currently favours. Thus it swung it in 1965 to educational programmes, and in 1968 swung it over to employment schemes.

The 1964 Act required CAP projects to finance 10 per cent of their costs locally in cash, kind or donated services.

In 1967 the local share was raised to 20 per cent. In the indefinite future the federal share is supposed to wither away entirely. This will probably never happen. More likely, America now has a permanent federally aided extension to its local welfare services. Even the termination of OEO need not disrupt this pattern, since many delegate agencies have now gained support from other federal departments, principally HEW, Labor, and Housing and Urban Development (HUD).

The rapid taming of community action was predicted by several of its more militant exponents. To Saul Allinsky, veteran protest group organizer from Chicago, CAP was 'a green wave of dollars, killing these militant, independent leaders of the poor, buying them off and killing them'. He himself accepted OEO funds in a controversial Syracuse university training scheme, claiming that he had to mop up some of this money to save it from drowning others. There are several independent community action groups, such as the church-run CAPs in Portland, Oregon, and the Black American enterprise, 'The Way' in Minneapolis, which, though modelled on OEO organizations, prefer to keep clear of federal funds.

The evolution of six typical CAAs in the San Francisco area has been outlined in an article in *Social Work* (Kramer and Denton, October 1967). The first people to see CAP opportunities were the welfare professionals, particularly those active on their local health and welfare councils. They tended to see CAP as a source of funds for existing agency objectives, although they rarely had any specific projects in mind. It was easy to legitimate themselves as the community's representatives. A meeting before an invited audience, called by the Health and Welfare council would elect the initiators as an action group. This would be progressively enlarged, in the later stages under

OEO pressure, without much loss of control by the initiators. Eventually it would secure recognition and a small federal planning grant, still without any specific projects.

Usually the CAA consists of a small policy board, and a large advisory board. An alert agency that lobbied for a place on the policy board in the formative stages usually won the right to nominate a member, creating a 'slot' structure of agency representatives.

The initiators usually withdrew to an advisory role once a permanent structure was set up, staff were hired and the board began to canvass for projects. However, the influence of the professionals remained, including almost always the representative of at least one leading public service, usually public assistance. Thus CAAs were mostly moderately conservative from the outset, and any radical zeal lay with individual delegate agencies or salaried staff.

9
Communities in action

There were 972 CAAs in early 1969, and they were all different. This chapter describes three of them, to illustrate the range of activities organized under CAA leadership in a medium-sized city, a large metropolis and an isolated rural area. All the material was collected in the summer of 1968.

Seattle

Seattle rivals San Francisco as the most beautiful city in the USA. Both Pacific cities have hills, bays and soaring bridges, but only Seattle has wild mountains in view to east and west, and the snow-flanked dormant volcano, Mount Ranier, looming across the southern horizon like a monstrous ice-cream cone.

This is one of the richest cities in America, and one of its most hectic growth spots, profiting very directly from the Vietnam war. Here in an intense degree is the paradox of the sixties; as wealth grows, poverty becomes more visible. The 7 per cent black minority has arrived almost entirely in the last fifteen years, building up the familiar ghetto with a 7 per cent unemployment rate while the aircraft factories still fly in labour from Britain and

Sweden. What is this city doing in community action?

The CAA board is the 33 member Seattle-King County Economic Opportunity Board, serving a population of about one million (about the size of Birmingham). 11 are *ex officio* members, including the mayor and heads of the major government welfare organizations; 11 are chosen by the mayor from various community groups; and 11 are 'poor' people elected at public meetings in different parts of the city. The action programmes operate through 10 delegate agencies, 2 of which (the local school boards) combine in the same programme. We will deal with each of their activities in turn, before moving across the continent for a more cursory tour of Philadelphia.

HEAD START

> 'There are five-year-olds who don't even know how to hold a book or how to listen to a sentence with more than three words. When we teach one of these children these things, we aren't making him "employable". But we will have begun to supply the missing ingredient—the catalyst—so that our regular school system won't lose that child along the way. And with that child we will have set in motion a process which in our time can spell the end of poverty.'
>
> (*Sargent Shriver, OEO Director, 12 April 1965.*)

Inspired by a mixture of Freud and Oscar Lewis, the largest Seattle CAA programme (about a quarter of total expenditure) assaults the 'culture of poverty' in the infant years. The local Head Start programme is delegated to the city and county school boards and to two neighbourhood agencies which run day nursery schemes. What it gives the poor children of the metropolis is not actually a head start, but the same form of pre-school experience that most of the middle class already provide for their children in private or co-operative nursery schools.

This is a carefully canned programme, with federal guidelines specifying a 3½-hour school day, hot meals with a certain nutritional content, medical examinations (but not treatment) and minimum pupil-teacher ratios. Free play has its place, but there is considerable teacher-directed activity. This emphasizes verbal skills, through story reading and singing traditional American nursery rhymes. All over the USA little circles of black and Spanish-American toddlers acculturate to the words of 'half a pound of two-penny rice'. Artistic self-expression and 'enrichment visits' to the zoo, the harbour and other attractions outside the slum areas further stretch mental horizons. There is a slight amount of formal instruction in recognizing colours and counting.

'Family life is fundamental to the child's development —Parents should play an important role in developing policies; will work in the Centers and participate in the programs' (Guideline No. 6, *Parents are Needed*). The local programme tries to implement this with parents' associations, which occasionally have their own enrichment visits and 'family life education' classes. In addition each class of 15 to 20 children is supervised by a qualified teacher assisted by a paid aide and a volunteer. Either of the assistants can be a Head Start mother, and at least one assistant is always drawn from the local community. The volunteer is often a VISTA. One object of the programme is to pioneer this sort of participation in the school system at large.

This is the least controversial part of the War on Poverty, although some more radical workers despise it as an aberration, stunting more militant activities. However it shares the same handicap as the ESEA remedial courses, the effect wears off. By the age of eight the Head Start 'graduates' show no significant advantages in school performances over non-graduates from the same income levels. There is a 'Head Start Follow Through' tutoring programme in some elementary schools, but it is not yet widespread. Short-lived 'remedial' experiences are no alternative to a general improvement in the educational system, and formal education alone is easily swamped by a hostile environment. But then that is why community action has a 'multi-faceted' approach.

Seattle area Head Start enrols over 1,000 children, mostly four year olds, on a one-year course. The children go on to public kindergarten, before starting elementary school at six. In many cities without kindergarten provision, the programme serves five-year-olds. About two-thirds of the schemes in the nation are limited to six or eight week summer courses, immediately prior to the pupils' entry into full time school. The original programme in the summer of 1965 was on this basis but it has since evolved into a full year course in most large cities.

Like most other anti-poverty enterprises it tends to cream its target population. The Seattle programme reaches about 25 per cent of the eligible population, that is four year olds from families beneath the federal poverty line (originally $3,000 per annum for two people, plus $500 for each child, maximum of $6,000). It is the less competent families which respond least and have most difficulty in keeping their children in the programme. The national average contact rate is about 29 per cent.

The income limit denies one broadening experience to the children, the chance to mix with other social classes and sometimes even other races. The one exception in the Seattle area was on a small island in Puget Sound where there was only one classroom for all the community's preschoolers, forcing combined Head Start/fee-paying cooperative classes.

Most community action day nursery services now include Head Start activities to gain earmarked funds. Here the eligibility limit is more crippling, forcing working mothers to withdraw their children if their earnings rise. This sometimes prohibits further employment, since satisfactory private day care is expensive and scarce in slum areas. OEO has recently tended to tighten the income test in all its programmes to focus funds on the most deserving cases, with some unfortunate side-effects. In the Fiscal Year 1967-8 over 200,000 American children participated in year-round Head Start programmes and nearly half a million joined summer-only schemes.

LEGAL SERVICES FOR THE POOR

> 'Unknown and unasserted rights are no rights at all.'
> (*Attorney-General Nicholas Katzenbach.*)

This is a canned programme with a difference. Most of its activity throughout America is in day-to-day services to its clients, usually arranging bankruptcies and divorces, but its most prized achievements have been test cases that have altered society's treatment of the poor. This federally funded programme eagerly sues federally subsidized agencies and even federal departments. Its triumphs include the abolition of several state residence qualifications on public assistance, which will certainly lead to their universal abandonment, on the grounds of constitutional right to freedom of movement. The widespread 'Man in the House' law, which stopped assistance to a woman cohabiting with a man, even where no financial support was provided or legally enforceable, has been overthrown; 'midnight' raids have been outlawed; and various conditions attached to relief payments have been revoked. As the result of a Washington D.C. ruling, a tenant now has the right to withhold rent if the landlord fails to make proper repairs and maintenance, and evictions from dwellings in breach of the local building by-laws are being challenged.

This service is provided by small groups of salaried lawyers, working for non-profit agencies in poor areas. The Seattle scheme was established by the Seattle Bar Association, and is controlled by a board of 15 trustees, including 7 representatives of the poor. Since 1967, by federal law all OEO schemes have been confined to civil cases. Criminal cases must resort to the legal aid schemes, which vary in different localities. These are usually supported from charitable funds, although Seattle is unique in that legal aid is supported on contributions from local law firms. Legal Services for the Poor in Seattle (1968) employed 11 full-time lawyers, and several law students as law-aides under the Work-Study programme.

A further activity of many legal services including Seattle is public education, with consumer groups and lecture courses. In July 1968 there were 266 Legal Services programmes in operation in America.

JUSTICE COURT PROBATION

This is a small, local, very dedicated enterprise to introduce a probation system into the magistrates courts of the city (it had previously operated only in the higher courts). The Courts are the delegate agency. It cheerfully accepts that most of its clients are not in officially-defined poverty, claiming it prevents poverty by stopping people going to prison. It seems fairly successful in getting probation accepted by the magistrates. At the time of my visit it had turned its attention to reforming the bail system. The office employed 5 probation officers and 2 aides in 1968.

OPPORTUNITIES INDUSTRIALIZATION CENTERS

This nationwide chain of training centres originated in Philadelphia as a black self-help project, and although OICs serve all races, they are usually managed by black Americans urgently reaching out to the hard-core unemployed of the ghettos.

Each centre offers an intensive training scheme, mostly in factory and office skills. Its distinctive feature is the 'feeder programs', taking entrants at any level and teaching basic skills and attitudes, if necessary re-cycling a trainee repeatedly, until he is fit for vocational training. The 'feeders' stress grooming, good speech, literacy and self-confidence. OICs are typically hung with brave slogans to 'instil motivation, encouragement and inspiration'. Earnest young black girls clack away at their keyboards behind large notices, 'I AM LEARNING HOW TO TYPE'.

The programme is jointly funded by OEO, HEW and the Department of Labor, plus local funds. It seems successful in reaching individuals who might not respond to the more conventional schemes. By the end of 1967 Seattle had 6 centres with 400 trainees.

NEW CAREERS

One of the latest canned programmes, this scheme developed from the book *New Careers for the Poor* (Pearl and Reissman, 1965) which argues that to escape poverty people need careers, not just recession-prone jobs. The greatest labour shortages are not in manufacturing industries but in the social services, which suffer accordingly. However, few of the unemployed poor have the formal educational qualifications to enter this work. New Careers attempts to create para-professional openings, with opportunities to reach professional rank, to relieve the manpower crisis, employ the poor and make the social services more effective, not only by enlarging them, but by making them more responsive to the needs and attitudes of their clients. It thus moves to break down worker-client distinctions, and effect cultural and institutional change simultaneously.

Most para-professional openings so far created are in the education system, including Head Start, principally as teacher aides. Teacher aides' duties range 'from supervision of recess and lunch-time activities to operating audio-visual equipment' (*New Careers for the Poor, Chapter 4*). For instance they assist children with homework, assume teachers' clerical functions, maintain supplies and special equipment, and exercise control over a class while the teacher gives special instruction to a particular child. The successful aides progress via in-service training to become teacher assistants, working in the classroom under the teachers' supervision when the class is divided into several groups for separate activities. The assistant is paid about half the salary of a teacher. Ultimately it is intended that assistants can progress through Associate rank and full-time training on secondment to become certified teachers.

Other para-professional activity is less clearly mapped. At the time of my visit the Seattle scheme had only recently started operation, and had made 77 placements, spread among the school system, public health, public assistance, employment services and the state and city recreation departments. The agency, multi-funded like OIC, arranges an initial training course through the local community college, and then finds trainee placements in local services.

EXTENDED SERVICES TO THE ELDERLY

This is a 'local initiative' programme whose main achievement has been to run a demonstration 'household aide' (home help) scheme, which stimulated similar projects in the public assistance and public health services. This venture had closed at the time of my visit, and the agency had become entirely a communications service, reaching out to the elderly on behalf of other agencies. The household aides had themselves been elderly poor, and most had used the experience to go on to related employment.

NEIGHBORHOOD HOUSE

An old-established settlement house with branch establishments in several public housing estates.

CARITAS

A straightforward tutoring scheme for 'slow learners', juvenile and adult, using 400 volunteer tutors, mostly university students.

CENTRAL AREA MOTIVATION PROGRAM (CAMP)

Seattle's black ghetto is an area of picturesque wooden houses in the 'carpenter gothic' style, put up during the boom days of the Alaska gold rush. It is loosely termed the central area, being close behind the commercial heart of the city, and is about two miles across with about 40,000 inhabitants. CAMP is a multi-purpose community association based on this district. When I was last there in July 1968, it was operating out of a converted fire station, ironically because its former H.Q. had been destroyed by arson. It is headed by a broadly based citizen's committee, but it is overwhelmingly a black venture, set up in 1965 under the sponsorship of the Urban League (a nationwide association for the advancement of urban negroes). The leadership is partly non-resident middle class, and the organization suffers somewhat from its identification as an anti-militant grouping. Every black American organization has to survive among a shoal of rival groups, competing to develop a new black role in society, and the policy of co-operation with the white establishment is currently widely out of fashion. The central area suffered heavily from arson last summer (1968) but there has been no looting or loss of life.

CAMP began work in late 1965 with a 'communications drive', recruiting 20 local people as 'block workers' (later renamed community organizers) for a doorstep canvas of the entire area. They were to introduce CAMP, and collect demographic data and information on local problems. Contact was made with most families, although a CAA survey two years later showed few of them remembered the event. Eventually the drive was abandoned as counter-productive, for it seemed to generate more suspicion than good will. Ever flexible, CAMP then concentrated its organizers on the most promising sector of the area, and set up 31 neighbourhood councils, each with 8 to 16 members drawn from a single block. These were to discuss local problems and seek solutions. Few of the councils had a very active life. There was most success where mutual aid networks were already strong, but in general this did not seem a very useful level for problem-solving, although some councils still survive.

Better success has attended the latest approach of area-wide issue-orientated groups, notably the 'motivated mothers'. This is an AFDC client group staffed by several CAMP personnel, which has affiliated to the National Welfare Rights Organization (NWRO). Several other groups have since formed outside the ghetto. The NWRO is one of the most effective client organizations in America, with an extremely stimulating influence on the public assistance system. The Seattle group has demonstrated outside the state capitol against a threatened cut in assistance grants, and had a TV confrontation with the state Governor. It has also made increased use of the appeals system in grant assessments, and has had a visible effect in improving the morale of its members, particularly the office holders. The group also functions as an effective family planning ('we've only had one casualty in the last six months') and consumer education group.

There is now also a separate Consumer Protection Group, organizing price surveys, a Tenants' Association, and an ad hoc Cross-Town Bus Committee, which secured a new bus route into the central area.

Within the communications section CAMP soon developed other types of employment. Ten community aides provide a domestic help service, while training to be community organizers. Ultimately these may be fed into OIC or New Careers, to make a permanent escape from poverty. (Community action is quite effective in making individuals mobile within the established system, even if it is difficult to influence the system itself.) Specialized youth workers have developed a Teen Club and various recreational activities, and an experiment with two 'detached workers' tries to contact non-involved youth. A neighbourhood newspaper *The Trumpet* started in 1966. Finally the section ran a housing survey, leading to a scheme to buy and rehabilitate old houses for sale to low-income families.

The other initial component of CAMP was 'Education-Guidance'. This consisted of an after-school tutorial programme based on study centres with VISTAs as 'out-reach workers', emphasizing remedial reading, and a youth vocational guidance service. Both enterprises were abortive. Tutoring related badly with school instruction, and the CAMP education aides have now moved inside the schools to teach 'Afro-American History'. This subject has appeared suddenly throughout all the educational media in America, attempting to add a positive black contribution to the usual version of American history. Since history is taught in American schools principally to cultivate national pride and identity, it is logical to re-vamp the curriculum when a more inter-racial self-identity is preferred. (This has led American Indians to demand a further revision of the textbooks. So far the British immigrants in Seattle—about 14,000—have not asked to re-write the Revolutionary War.) The Youth guidance service has now shifted to adult career training, turning the study centres into Action Education Centers to prepare more education workers.

Other ventures are now legion; 4 day-care centres with Head Start programmes, with their own bus service; a credit union; a drama group; a 'Beautification Project', now a separate agency on Department of Labor funds, using 42 hard-core unemployed on tree-planting work (one of its earlier achievements was to build a playground with 'the world's largest sandbox'): 'Soul Search', a high school and university exploration of the Black identity; a petrol station to train garage mechanics; and 5 'outreach stations'. These last, run by residents and VISTAs offer classes in cooking and handicrafts, tutoring, housing and employment counselling, and medical examinations by volunteer doctors and nurses. A family planning clinic failed through lack of clients. In summer there are several youth programmes, and the organization is now closely involved with the Model Cities project. The latter moves it away from dependence in OEO to other forms of federal support.

At the end of 1967 there were about 150 CAMP employees, of whom 116 were former low-income residents of the Area. In all King County there were about 800 paid CAP personnel, about 300 drawn from the low-income population.

In a personal interview the CAMP programme director told me that 70 per cent of the Area residents had been reached by his organization. However the CAA evaluation survey (June-July 1967) using a sample of 520 families, indicated only 30 per cent of the residents knew of CAMP and only 11 per cent felt they knew very much. Also 60 per cent of the persons contacted by CAMP were not low-income, tending to be better educated, more often employed and less likely to be elderly or disabled than the average. The reason lay partly in extensive neglect of the white poor, and in the allocation of blocks for intensive organization. However, CAMP moved in where it was most appreciated, not where the population was most apathetic.

The study centres (homework and tutoring) had the best support, reaching 37 per cent of eligible families, with youth activities reaching 13 per cent and job counselling 11 per cent. Communications workers had reached 22 per cent of all families (to the families' recollection) and

TABLE III

Extent of Various Federal Anti-Poverty Programmes in Three Cities, 1966-7, in Relation to Local Needs.

Manpower	*Detroit* (*Mich.*)	*Newark* (*N.J.*)	*New Haven* (*Conn.*)
Job training opportunities as % of unemployed population	Less than 50	Less than 20	Less than 33
Education			
% of eligible pupils reached.			
ESEA (Compensatory Education)	31	72	40
Adult Basic Education	2	6	4
Housing			
Housing units assisted as % of			
(i) Sub-standard units	2	16	14
(ii) Over-crowded units	1·7	23	20
Public Assistance			
Categorical assistance Persons served as % of poverty population	19	54	40
Community Action			
Persons reached as % of poverty population	30	44	42

Source: Report of National Advisory Commission on Civil Disorders, 1968

actively involved (e.g. on neighbourhood councils) 9 per cent. Head Start reached 25 per cent of eligible families in the sample, OIC 12 per cent and Day Care 2 per cent. These rates are similar to the Kerner Report's figures (*Report of the National Advisory Commission of Civil Disorder, 1968*) from Detroit (Mich.), Newark (N.J.) and New Haven (Conn.) (see Table III).

The experience of Community Action is of limited impact, mainly upon already socially active elements living in poor areas. Its varied achievements have been undoubtedly useful, and it has given many individuals a chance of social mobility. However, it does not seem, in its admittedly short history, yet to have had any serious engagement with the hard core of the 'culture of poverty'.

Philadelphia

Almost three thousand miles east of Seattle stands Philadelphia, the 'birthplace of independence', a city almost two hundred years old when the first six adults and six children landed at Seattle in 1851. It is still a very 'European' city, built of red brick rather than wood, with elegant eighteenth-century town houses and grim nineteenth-century slums, both faithfully copied from British models. Yet no Victorian slum ever stifled in such a humid 95 degrees as settles over Philadelphia in June, bringing swarms of flies into the narrow refuse-strewn courts.

Metropolitan Philadelphia has about 4 million inhabitants, but half of them live outside the city administrative boundary, and beyond the ambit of the city CAA, the Philadelphia Anti-Poverty Action Committee (PAAC). Within the city limits, 30 per cent of the population is black and there is a growing Puerto Rican minority. On present trends, the blacks will form the majority of the population in just over 10 years.

Philadelphia launched one of the nation's first community action ventures in 1961, on a Ford Foundation grant. In 1965 the programme was reconstituted, and recognized as a CAA. The constitution of PAAC is more elaborately defined than that of most CAAs. The governing board consists of 5 nominees of the mayor, although the mayor himself is not a member (in most city CAAs the mayor is a member, and in several cities, such as Chicago, he is chairman) and a county court representative; 12 representatives chosen from community agencies nominated by the mayor; and 12 representatives of the poor. Each of these last is elected by one of the 12 Community Action Councils (CACs). Each CAC has 12 members, and is elected by a geographically defined 'pocket of poverty'. About a third of a million people live in these 12 'pockets', 85 per cent of them blacks. All residents can vote in CAC elections (about 3 to 5 per cent do), but CAC members must not only be area residents but be living below the OEO poverty line. Also they may not be clergymen.

The CACs, besides being electoral colleges, act as neighbourhood councils, airing grievances and formulating proposals to submit to PAAC. Each also runs a local office, offering information and referral services like a British Citizens' Advice Bureau. Each office also has a full-time lawyer in attendance, as part of the *Legal Services* programme delegated to the Philadelphia Bar Association.

The other principal action programmes operate through 10 delegate agencies, although over 60 agencies are involved in the massive summer programme. This latter is a complex of education, camping and play activities (play streets, playgrounds and mobile pools), in which the leading agencies are the city parks department, the

Catholic archdiocese, the YMCA and the School District. About a quarter of a million people, mostly children and teenagers, are estimated to use the summer programme, which is very welcome in the oppressive summer climate. It also possibly serves to divert the poor from some other pastimes of a long hot summer.

Up to 1968 Philadelphia had a much stronger emphasis on education than in most city community action programmes. 62 per cent of the 1966-7 PAAC budget went to educational activities. Foremost was *Operation Get Set*, delegated to the School District, a day care and pre-school service for three and four year olds at 100 centres. This served 5,000 children on a year-round basis, employing nearly 1,000 staff, mostly ex-low-income. In contrast *Head Start* was an eight to ten week summer course for five year olds, also taking 5,000 children and delegated to the School District. The Philadelphia Association for Retarded Children ran a Head Start course for 30 children with IQs below 50. The Archdiocese (representing mainly the Italian, Irish and Puerto Rican minorities) ran *Operation Outreach*, an after-school study centre and tutoring programme, and *Operation Discovery*, a summer cultural course for junior secondary school pupils.

Philadelphia is the home of the original *OIC*, set up in a converted prison in 1964. There are now 4 Philadelphian centres with nearly 400 trainees. The founder was the Rev. Leon Sullivan, a six-foot-five-inch white-skinned Black American migrant from West Virginia, pastor of the Zion Baptist church and a fervent apostle of black self-help—'I see my ministry not . . . to give them milk and honey in heaven, but to give them ham and eggs on earth'.

As well as the OICs, which now have nearly 60 branches in the USA, Sullivan launched Zion Investment Associates, Inc., whose initial capital was raised by a small savings scheme among the Zion parishioners. This now owns a housing development (rented to middle-income tenants), a shopping centre and a clothing manufacturing company. This was followed in 1968 by *Progress Aerospace Enterprises*, set up with an investment from Zion, a bank loan, a training grant from the Department of Labor and technical advice from the General Electric Company. PAE is under contract with GEC, supplying electrical equipment for its Missile and Space Division. Like other Zion enterprises, PAE recruits labour mainly through the OICs. 150 employees were projected for March 1969. 40 per cent of the profits are reserved for neighbourhood education, health and welfare benefits and the rest is distributed to the workers and Zion stockholders.

Neither OEO nor PAAC has a direct stake in this venture in 'black capitalism', but in Columbus, Ohio, OEO has directly funded a ghetto co-operative to buy up small absentee-owned businesses. Following the 1967 disturbances, the owners are often eager to sell. The Community Self Determination Act now before Congress aims to give further federal support for such developments.

In 1966 OEO began to sponsor *Housing Development Corporations* to cultivate a significant private non-profit sector in the housing market. So far 11 corporations have been funded, and about 10 other private corporations exist. Their main emphasis is in buying and rehabilitating slum dwellings, and selling them for owner occupation by poor families, often to the existing tenants. We saw CAMP beginning this activity in Seattle. Philiadelphia has a quasi-public HDC, set up by the city in 1966, which acts as a PAAC delegate agency. The PAAC grant covers running costs and technical help, while the city provides the revolving fund, constantly being recovered by selling new or remodelled dwellings, and being reinvested in new projects. PHDC has so far completed about 100 dwellings, including about 20 new constructions.

So far HDCs have contributed less than 10,000 dwellings to America's housing stock, but they have demonstrated their feasibility principally in extending owner occupation to new social strata. PHDC's dwellings are transferred to neighbourhood housing associations, for sale or letting. The ultimate purchasers can benefit from the FHA below market interest mortgage scheme introduced in 1962, or the 1965 rent subsidies for non-profit landlords.

Another demonstration scheme in Philadelphia is the *Comprehensive Neighborhood Health Services Program* delegated to Temple University. This is based on 2 neighbourhood health centres. Ideally such centres should combine all non-hospital medical services, but initially Philadelphia is starting with dental and mental health. As in Britain it is difficult to get doctors into health centres, and a Seattle scheme as part of the Model Cities Program had to be replaced by a proposal for a patient reimbursement and travel expenses plan. The Philadelphia centres are also to possess dispensaries, and a health education component. There were about 30 OEO funded health centres in the USA in July 1968.

Foster Grandparents is another canned programme at work in Philadelphia (there were 66 such schemes in the USA in July 1968) delegated to the Health and Welfare Council. This employs about 80 elderly poor people with an interest in children to 'adopt' one or two children each from a local institution, often for the emotionally disturbed or mentally handicapped, and read to them, play games, take them on outings and give general loving attention. For 2 children, employees are expected to work 20 hours a week at $1·60 an hour. There is an initial two-week training course.

The latest delegated programme is *Consilio*, run by the newly formed Council of Spanish Speaking Organizations. This acts rather like a CAC for the Puerto Rican minority.

The Tulalip tribes, Washington

We have sampled community action in a moderately large city and in a great metropolis. Our last example is a small rural programme, in an unusual setting, an Indian reservation.

The Tulalip are a group of small fishing tribes in the Pacific North-West, brought together arbitrarily in a reservation set up just north of Seattle in the 1850s. Over the course of a century, most of the reservation land was sold off to non-Indians, leaving only a thin scatter of Indian homesteads. This process has now ceased and Indian land sales today are usually from individuals to the tribe collectively. In many ways this is a typical declining rural community, with the dispersed settlement common to much of America; but it is also a community trying to recover a tribal identity before it vanishes forever.

Community action in America began as a city enterprise, in the 'grey areas' identified by the Ford Foundation in 1960. The CAP effort tries to cover the nation but it is still strongest in the cities. The 30 per cent of the poor not in areas covered by a CAA in 1968 were mostly in rural locations. 40 per cent of the poverty population is rural, including the most extreme poor, but it absorbs only 30 per cent of the CAP budget. Rural programmes have attracted more attention recently, particularly with the OEO emergency food relief programme, but they remain an urgent area for expansion.

The Indian reservations see more community action than most rural areas, with nearly 200 Indian programmes by July 1968. Under the 1964 Act an elected tribal council can be a recognized CAA, so that no new organization is necessary. Like many rural CAAs the community action office of the tribal council is often little more than an outreach and referral service, and such further programmes as are developed come directly from the CAA, for there are few potential delegate agencies.

The Tulalip CAA was modest in its aims: 'To work with the people in any way possible, so they could get benefits which they were not getting or only getting a part of, and then only with a great deal of running around and a lot of waiting and red tape' (Report, March 1968).

One of the main achievements of rural CAAs has been to recruit clients for other services, especially OEO services. Thus the Tulalip CAA locates and enrols children for Head Start, and, more important, provides transport to often distant centres. In addition it recruited 10 youths for NYC in its first year's work, and 2 students for the local Community College; found employment openings for 31 people; recruited children for summer camps; listed families for Salvation Army Christmas parcels; recruited for Indian boarding schools (a traditional service of the Bureau of Indian Affairs); and canvassed for the federal welfare food programmes, and medical and financial aid services.

Economic development is a fundamental part of rural work, for instance in the development of agricultural co-operatives. In 1968 Tulalip development was still in the planning stage, aimed at checking out-migration by expanding tourism, fishing and basic services. One of the latest OEO earmarked rural programmes reserves funds for 'high impact areas', to check the drift from the land, and drain the reservoir of poverty that feeds the city slums.

CAP funds also helped expand the Tulalip's house

improvement scheme, hiring Indians for work and training on the project. The office also offered some legal services and started the tribal newspaper, *The Last Arrow*. The final paragraph of the brief mimeographed CAA annual report typifies the diversity and informality of its services.

> 'We have worked with 122 people who have either asked for help of one form or another, or we have offered to help.
>
> Some things which should be mentioned briefly are: Clothing for children so they could go to school. Bedding and wood for an older woman who was sick and had no heat. Enrolling children. Supplies for some who do what they can to earn money at home. Help with getting a college scholarship. Insurance problems.
>
> Please keep in mind that the program is here to help in any way possible. You are welcome to call on us at any time. The Office number is xxx-xxxx. My home number is xxx-xxxx.'

Above all this is an Indian programme, copying from Black America the attempt to base community action on a sense of racial pride and self-assertion.

> 'We shall always appear "different", because the majority of people in this land are not Indian . . . we wear the badge of our difference in such visible things as our hair, eyes and the tone of our skin. For this reason we cannot by forgetting our identity expect equality.'
>
> (*The Last Arrow*, editorial, April 1968.)

TABLE IV

% of Target Populations Covered by Various OEO Programmes, based on national sample surveys

Year	Programme	%
1968	Head Start	29
1969	CAP Parent and Child Centers	1
1965	CAP Migrant Day Care	2
1968	Upward Bound	4
1969	Neighborhood Youth Corps	6

Source: General Accounting Office evaluation study, 18 March 1969.

10
Fortunes of war

The American achievement

In July 1968 Acting OEO Director Bertrand Harding reported to the President:

> 'In the first 3 years of the OEO programs, from December 1964 to December 1967, 7 million Americans escaped from poverty status compared with 4·5 million in the previous 5-year period . . . Thus people in America have been moving out of poverty at 2½ times the annual rate of the previous 5 years . . . white Americans in poverty declined from one out of 5 in 1959 to one in 7 in 1964, to one in 10 in 1967. Between 1964 and 1967 non-whites came out of poverty 9 times as fast as during the previous 5 years.'

How much was attributable to federal anti-poverty efforts? Probably very little. Mollie Orshansky, writing in the *Social Security Bulletin*, earlier in the year (Orshansky, 1968), claimed most of the improvement occurred among the able-bodied poor, in or available for work. Full employment, to which the federal government's main contribution was the Vietnam war, was the most potent poverty killer. Employment and training programmes helped, and improvements in social insurance and assistance benefits also had some influence, although the major improvement in OASDHI did not take effect until early 1968. ('In a single stroke more than one million Americans were lifted above the poverty line'—President Johnson, March 1968.) It can truthfully be said that at the end of the Kennedy-Johnson administration, at least 12 per cent of Americans were still in poverty, and most of the improvement over the decade (from 24 per cent in 1959) was not directly attributable to federal domestic policy. Also, this improvement is measured on a static poverty line (in real terms), so that the relative deprivation of the remaining poor was much greater by 1969.

Yet this verdict alone is unfair. The Kennedy-Johnson regime saw the first important new departures in social welfare since the 1930s which opened the way for considerable further developments. Medicare is very limited in itself, but it is the beginning of a national health insurance scheme which will probably cover at least all OASDHI recipients within the next decade. Public assistance has seen important changes, and the new (1969) Secretary for HEW has endorsed his predecessor's call for 'national welfare standards', equalizing assistance payments throughout the nation. Like Medicare, this may take years of pressure to achieve, but the pressure continues to build. The Education Act broke the taboo on direct federal aid to schools, opening a route which will be more fully exploited. Racial discrimination survives, but all its outward forms have been outlawed, and the nation is perhaps ready for a little positive discrimination. The Nixon administration is not reformist, but there are important reforms still to be digested with potentially powerful results.

Also, much of the anti-poverty effort has been in education, particularly at the youngest ages. So far the results have been disappointing, but, as the improvements of educational standards advances beyond the provision of unsupported short-term remedial courses, it should yield positive returns. This is a very long-term measure. Even Sargent Shriver, OEOs first director, only promised an end to poverty 'in our time'.

But what of OEO? Have none of its enterprises yet left a visible mark on American life? It has scattered many useful little projects around the nation. About 100,000 former poor people are at work in community action enterprises, 75,000 of them in Head Start. A few thousand clients have also used OEO schemes to become socially mobile. Above all it has been a lovely war for the professionals such as myself—how else could I get high fees to visit the richest nation on earth to lecture on poverty? Yet this is trivial for a vast nation; what can one expect for less than 2 billion dollars?

The answer is that the War on Poverty was never intended as a 'first line' welfare service. It is a deliberately small-scale enterprise (although it is possibly too small to be fully effective), designed to trigger off changes in its wider environment; awakening the apathetic poor, re-educating insensitive bureaucracies, and incubating new types of service. It is by these repercussions that it must be judged.

The impact on the poor is difficult to assess. Contact rates by OEO enterprises have been quite low, reaching mainly the younger, healthier, better educated and better employed sectors of the poor. We have seen this in our Seattle example. Kramer records a similar situation in the San Francisco area (Kramer, 1969). The Kerner report suggests this in the low contact rates noted from three other cities (see Table III). Yet perhaps we do not need to reach the lowest and most unresponsive strata to start off a cultural revolution. As the Kerner report illustrated, the rioters of 1967 were not the most depressed ghetto residents. Undoubtedly some elements of the poor have become markedly more articulate, enterprising and demanding in the 1960s, with great consequences for the institutional structure of society, and thus ultimately for the mass of the poor.

Quite how much of the swirl of discontent in the late sixties can be attributed to OEO cannot be judged. It helped form many client organizations, and it may have provoked unfulfilled expectations in the ghettos that contributed to the epidemic of urban riots. Alternatively it may have directed some militant elements on to less destructive activities. With nearly a thousand CAAs the influence obviously differed between localities. Yet the War on Poverty has been more part of the general discontent than a force to cause or control it. It stirs or calms where it can, to focus discontent on chosen institutional targets.

The impact on the institutional structure is easier to see. It is felt in three ways: through external pressure by client groups, either by protest or advice, often with the help of OEO legal services; through infiltration from the ranks of the poor; and by internal reform prompted by the example of CAA agencies. Curiously the traditionally least progressive institutions seem to have responded most readily, perhaps because their morale was easier to undermine. An independent survey in 50 cities reported by OEO in March 1969 asserted the CAAs had played a major role in achieving significant institutional reforms in two-thirds of the sample communities. CAA influence was most effective in public assistance and employment services, much less so in the school system and voluntary welfare.

The National Welfare Rights Organization is probably

the most effective poor people's client organization in America. Although not an OEO creation, it has been promoted locally by many CAAs. It has helped feed test cases for OEO legal services which in four years have struck down time-honoured traditions of residence qualifications, maximum grants and midnight raids.

Many assistance departments offered no serious resistance to this pressure, and much of the personnel obviously welcomes it. One assistance office in suburban Philadelphia even had an NWRO stand in its waiting room piled with colourful literature under a banner 'Know your Rights'. Increasingly, departments are following the CAA style, with client and community participation, recruitment of aides from the poor (sometimes their own clients), use of volunteers, decentralization and 'outreach'. This last is the most difficult, since most state legislatures will accept that the assistance service should be less unpleasant to its clients, and even make itself more convenient to use, but actually looking for customers is seldom encouraged. However, CAAs, VISTAs and client groups continually flush out more and more eligible applicants, keeping the caseload mounting, despite increasing national prosperity and all the growth of preventive work.

Decentralization has mostly occurred through participation in 'multi-service centers' in slum areas. Many of these have been set up on CAA initiative, while in other localities, such as Seattle and Philadelphia, they have been provided by state or city authorities, following the OEO style. In Seattle the centre was offered in direct response to the threat of black violence in the central area in 1967. Public assistance and employment services are usually the core components of such centres, which may embrace a variety of other services, including recreational and social facilities. The move from the city centre office to the outreach station usually shifts the emphasis of the employment service from being a service to employers to becoming a personal placement service for the unemployed.

The school system has been the major area to employ poor people as para-professionals, most of them teacher aides passed on from Head Start. Many schools now possess such auxiliaries, which permits a more flexible teaching system. The classroom style will be even more influenced if and when greater numbers of aides become teachers' assistants. The use of volunteers is also spreading, mainly as after-school tutors. CAAs have also promoted curriculum changes, notably fostering the spread of Afro-American studies, and they have attempted to generate greater involvement between school and community in poor districts.

The ESEA programme reinforced this effort by requiring 'appropriate organizational arrangements' for community participation (July 1968 programmes guide) in all future projects, on the Head Start model. This has led to the creation of local advisory committees, but their influence on school affairs has yet to be felt. Community demands for a voice in the school system often go far beyond what administrators, and particularly teachers, will tolerate. This has been a major ingredient in the prolonged New York schools dispute.

The universities are clearly the most publicized example of the client revolt—not that American universities are particularly repressive, or that many of the students are underprivileged in the OEO sense. But then the school system would seem a less likely area of bitter conflict than the punitive public assistance service and its impoverished clients. One of the many paradoxes of the War on Poverty is that the least underprivileged have demanded their 'rights' most forcibly, and that the least repressive institutions have opposed these demands most fiercely.

OEO's direct influence on higher education has been limited to the small Upward Bound scheme, and counselling programmes in many of the Community Colleges. The former has brought a few more low-income, mainly black, students to various campuses, some of whom have become militants. It has also given greater scope for certain staff to experiment with new educational approaches. But also many students join VISTA, often as an interlude in their university careers. Others are drawn into CAA projects, particularly as tutors for 'underprivileged' children. The disputes in the school system, including disorder among senior pupils, cannot but be noticed by students who have only recently left these schools. Many of the specific disputes in American universities have concerned issues raised in the War on Poverty, such as the introduction of Afro-American studies and their status in the curriculum, and the impact of university physical expansion in slum neighbourhoods.

Once again we cannot claim the War on Poverty as a direct cause or consequence of the Student Revolt. But along with Civil Rights and Black Power, arises from the same social and psychological environment, and all have a very close interaction, sharing common ideas and techniques.

Of all the government schemes reshaped in the Community Action image, probably the Urban Renewal programme has changed the most. Once it was known as 'the federal bulldozer', which subsidized local authorities to blitz away entire communities, often redeveloping the sites for completely different populations. In 1966 it re-emerged as the Model Cities Program. Operating in 150 cities this attempts the comprehensive redevelopment, physical and social, of 'demonstration neighborhoods', with a high degree of resident participation.

In Seattle, for instance, the 'demonstration neighborhood' was naturally CAMP's established pitch, the Central Area, and the director of CAMP quickly became director of the new scheme—in his words, 'a chance to continue CAMP's work on many times its present scale'. The machinery for citizen participation followed the complex CAA pattern. The Mayor appointed all the staff, and nominated 100 local organizations each to elect a representative to the Model Cities Advisory Council. This Council elected a Chairman, who nominated the chairmen of 6 Citizen Task Force Planning Committees (Health, Housing, Physical Planning, Education, Welfare and Employment). The Advisory Council then matched this with 6 co-chairmen, the 12 individuals becoming the Model Cities Steering Committee.

All the planning took place at the task force meetings, at week-ends or evenings in schools and social centres in the ghetto. Here the chairmen of each task force met with interested citizens, locals and outsiders. Decisions were made on the vote of the local residents only, even though these were sometimes a tiny minority of the meeting, participating little in the debate and giving no indication that they understood the motions proposed. Other meetings, however, were crowded and clamorous, particularly those of the education task force.

The federal government meets 80 per cent of the general costs of this programme, plus the full cost of re-locating uprooted households. Re-location is only an incidental effect of the scheme, which 'gilds the ghetto' rather than dispersing it. It even allows America's ghettos a crumb of self-government.

In February 1969 the new President gave his perception

of OEO. This is crucial since the 1964 Act expires in 1970, and President and Congress must decide if and how the War on Poverty is to be continued. Nixon described OEO as an 'incubator', warming up experimental projects. Evidently several of OEO's present clutch of eggs are now well hatched, for under his executive authority the President proposes to delegate Head Start to HEW, to form the nucleus of a new office of Child Development. Head Start funds are to be increased to convert most of the summer schemes into year-round programmes, and the Follow Through programme is to be stepped up. A re-shaped Job Corps is to be delegated to the Department of Labor, closing most of the Conservation Centers in favour of small urban or suburban camps, residential but recruiting from the locality. Eventually the President also hopes to shift Comprehensive Health Centers and Foster Grandparents to HEW, all 'to free OEO for a greater concentration of its energies on its innovative role'.

It is more humble to be called an 'incubator' than 'a total commitment . . . to pursue victory over the most ancient of mankind's enemies' (OEA presidential message, 1964), but it is still a vital role. The emphasis is now more clearly on services rather than organization, but this does not preclude a growing interest in functional poor people's organizations, such as co-operatives and community corporations. It seems that OEO will survive to hatch more offspring and maintain the dynamic that has at last entered America's social services. If the 1970s are at all like the previous decade, they will need it.

Only in America?

Earlier we described the War on Poverty as a uniquely American enterprise, moulded by the creed of individual participation and self-help; the belief in poverty as a form of cultural lag bound up with racial, ethnic and regional differences; the anxiety to forge new relationships between target populations and the federal government born of repeated frustration with the old political and administrative arrangements; and the tolerance of conflict as socially useful to a degree unusual even among democracies. Rooted as it is in the American situation, does this make the experience irrelevant to other societies?

The industrial revolution was born out of the unique circumstances of eighteenth-century England, but it has not been irrelevant to other nations. America, for its own special reasons, was the first industrial nation to use community action as a domestic technique, and to reach so far in positive discrimination for target populations, but this certainly has a relevance elsewhere. Just as Britain followed three or four years after America in the literary, statistical and political rediscovery of poverty, so now it is beginning to pick up some American combat techniques. The Plowden Report (1967) recommended more community participation in British schools, the wider use of teachers' aides and assistants and the establishment of 'Educational Priority Areas', which like ESEA are to channel central government resources to schools in disadvantaged localities. The government set up the EPAs and followed on with a plan for expanded nursery education in the same districts. The minister for the social services gave a short TV lecture on the importance of early childhood experience that could well have been taken from a Head Start hand-out. Shortly afterwards a five-city pilot community action scheme was launched to experiment in community participation, in particular in the delivery of social services. And in July 1969 Balliol College, Oxford ran Britain's first Upward Bound programme.

America came to community action partly because its 'first-line' social services were underdeveloped. At first local initiative was a substitute for a first-line structure and later became a system to test out possible extensions to it. Most of Europe already has a well-developed 'first line', but this is a situation in which community action probably has its greatest potential. The better a nation's social services, the more valuable it is to have a strong outreach component.

Shortly before I left for America in 1966 two students and myself studied an excellent national health service hospital outpatients' service in Birmingham. The paediatric clinic in particular had a very high non-attendance rate for its appointments, concentrated among patients from the poorer districts of the city. These were the most distant part of the hospital's 'catchment area', badly served by public transport. A follow-up study of the non-attenders discovered a population of large families in which minor children's illnesses were perpetual, a high incidence of broken families, many families where the mothers and children spoke no English and a general situation where bringing the children to the clinic or getting baby sitters for the usual two hours' absence was daunting. Once an appointment was missed, some parents, especially immigrants, felt too compromised to attend again. Yet hardly anybody denied their children needed attention (E. James, 1966).

A local out-reach service might have been able to help with transport, interpreters, escorts or baby-sitters. Volunteers might have been useful in the clinic waiting-room minding children while their mothers saw the consultants or collected refreshments. A neighbourhood centre could re-establish contact with alienated patients, and might assist in planning for a local health centre. It might also provide a forum for patients to express their views and for the hospital to forge better public relations, helping it meet situations such as the Indian expectant mother who avoided the ante-natal clinic in terror of being admitted as an in-patient. Community action is no substitute for the National Health Service, but they can make each other far more effective.

British society is closer to that of America, and more responsive to American examples than the rest of Europe, yet France too has re-discovered poverty on a scale comparable to the USA. Europe is open to a new form of Americanization, not the once-familiar conservative chromium-plated complacency, but an anxious introspective unrest. The western world has passed the first intoxication of affluence. Those who in one way or another feel under-privileged—black people, poor people, students and juveniles—have become more articulate and demanding even as the strongest economic pressures on them have lifted. Majority sympathy can be translated into political support, but it is easily alienated. Societies of mass affluence have yet to prove that they can be made fully responsive to the needs of minorities, but they can no longer pretend that the needs do not exist.

Postscript – August 1969

On August 8th, 1969, President Nixon made his first major policy statement on social welfare, nearly eight months after taking office. This statement announced an internal re-organization of OEO, giving further emphasis to its research and development role, and upgrading the legal and health services divisions. The purpose of OEO was closely defined as 'the "R. and D." arm for government's social programs . . . the cutting edge by means of which government moves into unexplored areas'.

More far-reaching proposals, which will require Congressional assent, were made in revenue sharing, manpower training, and income maintenance. A 'block grant' system of federal support, similar to the financing of British local authorities from the central government, is proposed for state and local governments as well as the closely tied grants-in-aid of the past. The various manpower training schemes including the Job Corps are to be pulled together under a single act, and made the complete responsibility of the Department of Labor. Finally, a guaranteed federal minimum income is to be made available to the poor of every state, of $1,600 for a family of four (i.e. half the OEO poverty line). This 'cannot claim to provide comfort for a family of four, but the present low of $468 a year cannot claim to provide even the basic necessities'.

This federal mimimum is proposed for every poor family, with or without a father, and whether or not the father is employed. With employed persons a system of 'disregarded' income aims to maintain the financial incentive to work.

All this has yet to pass Congress. However it is significant that a Republican president has proposed a greater federal expenditure on welfare than either of his Democratic predecessors. America's commitment to fight poverty has survived the collapse of the Johnson administration, and OEO has survived as an important part of the anti-poverty effort.

Suggestions for further reading

American Civilization, An Introduction, edited by A. N. J. Den Hollander and S. Skard, Longmans, 1968, is a good overall introduction written for European readers by a group of scholars, most of them Europeans. S. M. Lipset's *The First New Nation*, Heinemann, 1963, is a stimulating social psychological study of America, with comparisons with Britain and Sweden. *This USA*, by B. Wattenberg and R. M. Seammon, Doubleday, 1965, is an excellent statistical survey, based on the 1960 Census.

The Swedish sociologist Alva Myrdal wrote the classic work on Black America, *An American Dilemma*, Harper & Bros., 1944. This was before the major black migration. The Kerner Report (*Report of the National Advisory Commission on Civil Disorders*, US Govt. Printing Office, 1968) contains an up-to-date survey of demographic and political movements among black Americans, as a background to the riots of 1967. This report also contains a survey of anti-poverty efforts in three major cities. The Moynihan Report (*The Negro Family and the Case for National Action*, US Dept. of Labor, 1965) is a gloomy statistical analysis of black American family life. Its controversial reception is described in L. Rainwater and W. L. Yancey's,

The Moynihan Report and the Politics of Controversy, MIT Press, 1967. This also contains the full text of the report.

Oscar Lewis' principal studies of the culture of poverty are contained in *Five Families*, Basic Books (New York), 1959, *Children of Sanchez*, Secker and Warburg (London), 1962 and *La Vida*, Secker and Warburg, 1967.

There are relatively few studies of 'normal' middle-class American life, in comparison with the abundant minority group studies. J. K. Galbraith's *The Affluent Society*, Hamish Hamilton (London), 1958, touches most aspects of American life, including poverty. The pioneer poverty study of the 1960s is Michael Harrington's *The Other America*, Penguin Books, 1963. One of the more recent surveys is *Poverty—American Style*, a very useful 'reader' edited by H. P. Miller, Wadsworth Publishing Co. (Belmont, California), 1966. This includes a summary of the contemporary welfare system, and the diet plan on which the federal poverty line is based. One of the few comparative studies in life styles is the recent US-British-Danish study of the elderly, E. Shanas and others, *Old People in Three Industrial Societies*, Routledge, 1968.

E. S. Griffiths, *The American System of Government*, Methuen, 3rd Edition 1962, gives a concise survey of American government for British readers. The main text-books on the American social services are E. M. Burns, *Social Security and Public Policy*, McGraw Hill (New York), 1956, C. A. Schottland, *The Social Security Program in the United States*, Appleton-Century Crofts, 1965, W. A. Friedlander, *Introduction to Social Welfare*, Prentice Hall (New York), 1968, and H. L. Wilensky and N. Lebeaux, *Industrial Society and Social Welfare*, The Free Press (New York), 1965. The US Welfare Administration has published a 47 page 'capsule view' of American welfare by F. A. Koestler, *Portrait of Median City*, Publication No. 16, US

Govt. Printing Office, 1966. The National Association of Social Workers periodically publishes a very useful reference work, *Encyclopaedia of Social Work*, latest edition 1965, editor H. L. Lurie.

Concerning the rediscovery of poverty, we have already noted Galbraith and Harrington. H. Gordon edited the proceedings of a national conference on poverty in which most of the leading authorities participated, *Poverty in America*, Chandler Publishing Co. (San Francisco), 1965.

The main work on malnutrition in America is *Hunger, USA*, by the Citizens' Board of Inquiry into Hunger and Malnutrition, New Community Press (Washington), 1968. The British 'poverty literature' is exemplified by Abel-Smith and P. Townsend, *The Poor and the Poorest*, Bell, 1965 and by the government survey, *Circumstances of Families*, HMSO, 1967. Britain's 'rediscovery of poverty' is outlined in B. Rodgers', *The Battle Against Poverty, Vol 2*, Routledge, 1969. France's equivalent of Michael Harrington has been P. M. de la Gorce, whose book *La France Pauvre*, Bernard Grasset (Paris), 1965, is dedicated to The Other France.

The Kennedy-Johnson administration is too recent for there to be any overall survey of its domestic legislation, information on which must be drawn mainly from the *Congressional Quarterly*. Occasionally this publishes supplements, including *Congress and The Nation*, 1965, which outlines all Congressional activity, 1945-64. *Social Problems: A Modern Approach*, edited by H. S. Becker, J. Wiley and Sons (New York), 1966, includes a chapter by S. M. Miller and M. Rein, 'Poverty, Inequality and Policy', which surveys the major anti-poverty strategies in use. C. Green, *Negative Taxes and the Poverty Problem*, Brookings Institution, 1967, surveys the main arguments on this topic.

SUGGESTIONS FOR FURTHER READING

At the beginning of 1969 there was very little literature directly on the work of OEO, except in periodicals. D. P. Moynihan has published a very partisan account of community action, *The Maximum Feasible Misunderstanding*, Free Press, 1969. The community action projects 1961-4 which foreshadowed CAP are described in P. Marris and M. Rein, *Dilemmas of Social Reform*, Routledge, 1967 which examines the contradiction between community action and community organization. This is examined again, four years later in California, in five case studies by R. M. Kramer in *Participation of the Poor*, Prentice Hall, 1969.

OEO itself publishes several periodicals, including a weekly newsletter, monthly accounts of the rural community action programme, the legal services programme and VISTA (*Rural Opportunities, Law in Action*, and *Vista*) and a bi-monthly magazine, *Communities in Action*. This is available from the Public Affairs branch of OEO, Executive Office of the President, Washington, D.C. Fact Sheets and other information are available on request.

Other relevant professional journals are *Social Work*, the journal of the National Association of Social Workers, *Public Welfare*, the journal of the American Public Welfare Association, and *Social Security Bulletin*, published by the Social Security administration. This last contains most of Mollie Orshansky's articles.

Bibliography

C.C.C. INC., *Program of Neighborhood Development and Organization*, Minneapolis, 1967.

GALBRAITH, J. K., *The Affluent Society*, Hamish Hamilton (London), 1958.

GORDON, H., *Poverty in America*, Chandler Publishing Co. (San Francisco), 1965.

HARRINGTON, M., *The Other America*, Collier-Macmillan (London), 1962. Also Penguin, 1963.

JAMES, E., 'Part-Time Hospital Patients', *Medical Care*, April-May, 1966.

KERNER REPORT, *National Advisory Commission on Civil Disorders*, US Govt. Printing Office, 1968.

KRAMER, R. M., *Participation of the Poor*, Prentice Hall, 1969.

KRAMER, R. M. and DENTON, C., 'Organization of a CAP', *Social Work*, October 1967.

LEWIS, O., *Five Families*, Basic Books (New York), 1959. *Children of Sanchez*, Secker and Warburg (London), 1962. Also Penguin.

MILLER, H. M., *Characteristics of Families receiving AFDC*, Bureau of Public Assistance, April 1963.

MOYNIHAN, D. P., *The Negro Family: The Case for National Action*, US Dept. of Labor, 1965.

ORSHANSKY, M., 'Counting the Poor: Another look at the poverty profile', *Social Security Bulletin*, 28(i), 1965.

'The Shape of Poverty in 1966', *Social Security Bulletin*, 31(iii), 1968.

PEARL, A. and REISMAN, T., *New Careers for the Poor*, Free Press, 1965.

PLOWDEN REPORT, *Children and their Primary Schools*, HMSO, 1967.

TAENBER, K. E. and A. F., *Negroes in Cities*, Aldin Publishing Co. 1965.